Entering Mentoring

A Seminar to Train
a New Generation of Scientists

Jo Handelsman
Christine Pfund
Sarah Miller Lauffer
Christine Maidl Pribbenow

THE WISCONSIN PROGRAM FOR SCIENTIFIC TEACHING
• Supported by the Howard Hughes Medical Institute Professors Program •

Entering Mentoring

Jo Handelsman

Christine Pfund

Sarah Miller Lauffer

Christine Maidl Pribbenow

•

THE WISCONSIN PROGRAM FOR SCIENTIFIC TEACHING

• *Supported by the Howard Hughes Medical Institute Professors Program* •

Contact Information:

Phone: (608) 265-0850

Email: scientificteaching@mailplus.wisc.edu

Website: http://scientificteaching.wisc.edu

The development of this book was supported by a grant from the Howard Hughes Medical Institute to the University of Wisconsin-Madison in support of Professor Jo Handelsman. HHMI

With contributions from:

Janet Branchaw, Center for Biology Education

Evelyn Fine, Women in Science and Engineering Leadership Institute

HHMI Graduate Teaching Fellows and their faculty mentors

Edited by:

Hilary Handelsman

Front Cover:

The cover art is a fractal image, entitled, "Fractal Mitosis," by Jay Jacobson, the creator of the art form, Fractalism™

For PDF version of this book, go to
www.hhmi.org/grants/pdf/labmanagement/entering_mentoring.pdf

Preface

Effective mentoring can be learned, but not taught. Good mentors discover their own objectives, methods, and style by mentoring. And mentoring. And mentoring some more. Most faculty learn to mentor by experimenting and analyzing success and failure, and many say that the process of developing an effective method of mentoring takes years. No two students are the same or develop along the same trajectory, so mentoring must be continually customized, adjusted, and redirected to meet each student's needs. A skilled mentor's decisions and actions are guided by a reflective philosophy, a well-developed style, and an ability to assess student needs. There is certainly no book that can tell us how to deal with every student or situation, but a systematic approach to analyzing and discussing mentoring may lead us to a method for tackling the knotty challenges inherent in the job.

The goal of the seminar outlined in this manual is to accelerate the process of learning to be a mentor. The seminar provides mentors with an intellectual framework to guide them, an opportunity to experiment with various methods, and a forum in which to solve mentoring dilemmas with the help of their peers. Discussing mentoring issues during the seminar provides every mentor with experience—direct or indirect—working with diverse students, tackling a range of mentoring challenges, and considering a myriad of possible solutions. Members of the seminar may hear about, and discuss, as many mentoring experiences as most of us handle in a decade, thereby benefiting from secondhand experience to learn more quickly. We hope that, when mentors complete the seminar, they will have articulated their personal style and philosophy of mentoring and have a toolbox of strategies they can use when faced with difficult mentoring situations.

The anticipated outcome of this seminar is twofold. First, we want to produce confident, effective mentors. Second, we intend this seminar to have a far-reaching effect on the undergraduate research experience. Undergraduates obtain numerous benefits from participating in independent research and those benefits can be amplified by good mentoring. Both outcomes enrich the research experience for everyone involved.

We developed the mentoring seminar presented in this manual as part of The Wisconsin Program for Scientific Teaching, using an iterative approach of developing, testing, evaluating, and revising our teaching methods and seminar content. The material that survived is the result of seven different seminar cohorts led by four different facilitators. Therefore, the seminar presented here has been tested in mixed formats by various facilitators. We have included topics that emerged repeatedly, questions that consistently generated discussion, and readings that were universally appreciated by the mentors.

Everyone who has taught this seminar has enjoyed it, and felt changed and enriched by it. Novices who have never run a lab or research group seem to be as effective at running this seminar as seasoned faculty with decades of mentoring experience. We assume that this is because the discussions are propelled by the participants, not the facilitator, and all of us can draw on our experiences as mentees, even if we don't have experience as research mentors. As long as the facilitator asks a few key questions, keeps the discussion focused, respectful, and inclusive, and helps the mentors see the patterns and principles raised in discussion, the seminar will be a success. We wish you fun and good mentoring as you embark on what we hope is a remarkable teaching and learning experience.

Jo Handelsman
Christine Pfund
Sarah Miller Lauffer
Christine Maidl Pribbenow

Contents

Mentoring Seminar Content, Format, & Implementation

Content

The content of each session is designed to address the key concerns and challenges identified by mentors we interviewed. The topics include:

- *intellectual issues:* comprehension and learning to ask questions

- *technical issues:* experimental design, precision, and accuracy

- *personal growth issues:* developing confidence, creativity, and independence

- *interpersonal issues:* dealing with students of diverse experiences and backgrounds, motivation, honesty between mentor and student, scientific integrity, and discrimination

Format

In the discussion sessions we facilitated, we used a very open discussion format. Simply asking the mentors a few guiding questions led to vigorous discussion. The case studies and reading materials often provided a tangible starting point, but the mentors quickly moved from the hypothetical examples to their own experiences with students. The seminar is most effective with mentors who are working with students full-time, as, for example, in an undergraduate research summer program, because the short duration of the program intensifies the urgency of dealing successfully with challenges that arise. Likewise, the frequent contact with the students provides mentors the opportunity to implement immediately ideas generated by the discussions.

Implementation

Prior to the start of each session, copy the white pages for each mentor in your group. These pages contain the materials (readings, guiding questions, etc.) mentors will need for the following session. Alternatively, all the white pages can be copied at the start of the seminar and distributed at the first meeting. Guiding questions and notes for group facilitators are pages printed on the light blue pages, and are arranged in the manual session by session. It is important to have the first meeting with mentors before the mentees begin work in the lab.

Big Questions in Mentoring

Below are some guiding questions that may be useful in discussions about mentoring.

Expectations

- What do you see as your student's greatest strength(s)?

- What area(s) do you think your student should focus on developing? How do you suggest they do this, and how can you facilitate this process?

- What do you expect your mentee to accomplish while in the lab?

- How independent should your mentee be?

- How much assistance do you expect to provide for your mentee?

- What do you hope to get out of the mentoring experience?

- What does your mentee hope to get out of the research experience?

- What have you learned about working with your student that you did not expect to learn?

Scientific Teaching

- What is your approach to mentoring?

- How does the concept of "Scientific Teaching" apply to mentoring?

- Does your approach to mentoring involve active learning strategies?

- What evidence do you have that your approach to mentoring is effective?

- What evidence would convince you that your approach to mentoring is effective?

- How could you improve on your mentoring based on student feedback?

Community of Resources

- What is the value of presenting mentoring challenges to your peers and hearing their approaches to a given challenge?

- Do you see your peers as a valuable resource in addressing mentoring issues?

- Do you see your adviser or another faculty member in your department as a resource on mentoring?

- Do you see your department as a network of mentors?

- How could you create a stronger community of mentors and mentoring resources?

Diversity

- How do you define diversity?

- Have you created an environment that allows your mentee to benefit from the diversity in your lab/department? How?

- How might another mentee with a different learning style, personal style, or background view your mentoring approach?

- How do you deal with diverse learning styles, personal styles, ethnicity, experience, gender, and nationality?

Syllabus for Mentoring Seminar

Dates	Topics	Assignments Due	Readings
Session 1	**Getting Started** • Introductions • The elements of a good research project • Establishing a good relationship with your mentee		"Teaching Scientists to Teach"; J. Handelsman "Scientific Teaching"; J. Handelsman *et al.*
Session 2	**Learning to Communicate** • Case study: projects • Mentees and their projects • Establishing expectations—the mentor's and the mentee's	1. A paragraph describing your mentee's project 2. Written mentoring philosophy	"What is a mentor?" in *Adviser, Teacher, Role Model, Friend;* NAS
Session 3	**Goals and Expectations** • Mentoring philosophies • Case study: trust • How do you know that they understand what you are saying?	1. A short biography of your mentee with information you gather from interviewing them. 2. Summary of the discussion you and your mentee had about expectations	"Mentoring: Learned, Not Taught"; J. Handelsman
Session 4	**Identifying Challenges & Issues** • Case studies from your first few weeks—challenges and suggestions • How do you know if there are problems?		
Session 5	**Resolving Challenges & Issues** • Proposed solutions to the issues raised in the case studies • Case studies: diversity • Midcourse process check	A written proposal of a possible solution to one of the challenges described during a previous mentoring discussion	"Benefits and Challenges of Diversity"; WISELI
Session 6	**Evaluating Our Progress as Mentors** • Mentoring challenges and suggestions	Thoughts about how you and your mentee differ. How do these differences affect the summer experience for both of you?	"Righting Writing"; J. Handelsman
Session 7	**The Elements of Good Mentoring** • What can we learn from other mentors? • What has proven effective in your mentoring? • Presentations	Present one of your mentoring challenges to your PI (or another you respect as a mentor) and ask how they would handle the situation. Submit a summary of their response and what you thought about it.	
Session 8	**Developing a Mentoring Philosophy** • Mentoring philosophies after the mentoring experience	Rewritten mentoring philosophy	

Mentoring Seminar: Session by Session

Facilitator Notes and Materials for Mentors

Session 1:
Getting Started

Discussion Outline: Session 1

Topics:

Introductions

Describe the Mentoring Seminar

Discuss Seminar Logistics
1. Syllabus
2. Assignments
3. Confidentiality

Discussion Questions
- What are the elements of a good research project?
- How can you establish a good relationship with your mentee?

Describe Assignments for Session 2: Mentoring Philosophies and Project Descriptions.

Materials for Mentors:

Syllabus

Goals of the Mentoring Seminar

Concepts, Techniques, and Practices to Teach Your New Mentee

Permission Form

Reading: "Scientific Teaching"

Reading: "Teaching Scientists to Teach"

Session 1:

Introductions

The Mentoring Seminar

This program is dedicated to improving the mentoring skills of a new generation of graduate students and postdoctoral researchers who may become science faculty. The seminar is designed to help them become effective mentors to diverse students using discussions, collective experiences, and readings to develop strategies for mentoring.

Syllabus

An eight-session syllabus is included. It is recommended that you meet for one hour each session with your group of mentors.

Assignments

To foster positive attitudes toward the assignments, it may be helpful to explain to your group of mentors that the assignments are meant to provide a framework for them to reflect on their roles as mentors and to encourage them to view their mentoring as an opportunity to engage in scientific teaching. Distributing student writing to the entire group can prompt interesting discussion.

Confidentiality

You will need to discuss confidentiality with your group. As a cohort, your group should agree on a policy for confidentiality regarding the ideas shared in each session. It is important that the group recognizes that regardless of its confidentiality policy, it cannot control everything that is said outside the room. If members have any concerns about sharing information, they should tailor their comments accordingly. In addition, members should decide if they would like their names removed from any writing assignments before they are compiled and distributed to the group. Last, The Wisconsin Program for Scientific Teaching asks each participant to consider signing a permission form regarding future use of their case studies. This form is included in this section.

What are the elements of a good research project?

Mentors usually have a strong sense of what constitutes a good research project. Ask them, as a group, to come up with the elements of a good research project. Some thoughts that have emerged in previous discussions are:

- Projects should have a reasonable scope
- Projects should be feasible
- Projects should generate data that the student can present
- Projects should not simply include cookbook experiments
- Projects should have built-in difficulties that will be faced after the student has developed some confidence
- Projects should be multifaceted

How can you establish a good relationship with your mentee?

One way to start this discussion is to ask the mentors what they should do the first time they meet with their mentee. Even if they have already met the student, this discussion can help the mentors consider the importance of the personal interaction they have with their mentee. Some thoughts from previous discussions are:

- Make direct eye contact
- Be enthusiastic
- Introduce them to the lab and your lab mates
- Acquaint them with the building
- Get them started on a lab notebook
- Talk about the "big picture"
- Discuss lab policies
- Discuss the mentee's background
- Get to know your mentee

Many mentors expressed concern that the undergraduate researchers with whom they were working either did not know basic lab protocols and techniques or needed to be reminded of them. One mentoring group developed a list of techniques and skills every undergraduate researcher should know. A copy of that list is included in this section. This particular list was developed by mentors working in molecular biology labs; mentors working in other fields may wish to create a similar list more relevant to their lab's area of inquiry.

Assignments

Ask mentors to write a paragraph describing their mentee's project.

Have each mentor describe their mentoring philosophy in writing. There is no length requirement. As the facilitator, you may wish to develop your own philosophy and share it with the group.

Dates	Topics	Assignments Due	Readings
Session 1	**Getting Started** • Introductions • The elements of a good research project • Establishing a good relationship with your mentee		"Teaching Scientists to Teach"; J. Handelsman "Scientific Teaching"; J. Handelsman *et al.*
Session 2	**Learning to Communicate** • Case study: projects • Mentees and their projects • Establishing expectations—the mentor's and the mentee's	1. A paragraph describing your mentee's project 2. Written mentoring philosophy	"What is a mentor?" in *Adviser, Teacher, Role Model, Friend*; NAS
Session 3	**Goals and Expectations** • Mentoring philosophies • Case study: trust • How do you know that they understand what you are saying?	1. A short biography of your mentee with information you gather from interviewing them. 2. Summary of the discussion you and your mentee had about expectations	"Mentoring: Learned, Not Taught"; J. Handelsman
Session 4	**Identifying Challenges & Issues** • Case studies from your first few weeks—challenges and suggestions • How do you know if there are problems?		
Session 5	**Resolving Challenges & Issues** • Proposed solutions to the issues raised in the case studies • Case studies: diversity • Midcourse process check	A written proposal of a possible solution to one of the challenges described during a previous mentoring discussion	"Benefits and Challenges of Diversity"; WISELI
Session 6	**Evaluating Our Progress as Mentors** • Mentoring challenges and suggestions	Thoughts about how you and your mentee differ. How do these differences affect the summer experience for both of you?	"Righting Writing"; J. Handelsman
Session 7	**The Elements of Good Mentoring** • What can we learn from other mentors? • What has proven effective in your mentoring? • Presentations	Present one of your mentoring challenges to your PI (or another you respect as a mentor) and ask how they would handle the situation. Submit a summary of their response and what you thought about it.	
Session 8	**Developing a Mentoring Philosophy** • Mentoring philosophies after the mentoring experience	Rewritten mentoring philosophy	

Goals of the Mentoring Seminar

The goals of the Mentoring Seminar are to guide mentors to:

understand "scientific teaching" and apply it in mentoring so they can:

- become more reflective and effective mentors, and
- design, implement, and evaluate various approaches to mentoring

build a relationship based on trust and respect with a mentee so they can:

- communicate more effectively with mentees

- stimulate creativity, independence, and confidence in mentees

- work well with students of diverse learning styles, personal styles, experiences, ethnicities, nationalities, and gender

build a community with other mentors by:

- sharing mentoring challenges and solutions with each other

Concepts, Techniques, and Practices to Review with Your New Mentee*

1. Remind them that it is better to ask questions than to make a mistake that could have easily been avoided.

2. General lab safety procedures including:
   a. Appropriate clothing
   b. Food and drink in the lab
   c. Lab coat/gloves/glasses

3. How to find and use helpful reference manuals such as Current Protocols

4. Chemical and biological safety issues including:
   a. How to dispose of wastes
   b. How and when to use a fume hood
   c. How to handle chemicals safely
   d. How to clean up a spill
   e. How to assess whether a particular chemical should be handled in a fume hood
   f. How to handle and dispose of biological materials

5. "Chemical hygiene"—cleaning up, discarding excess (not returning waste to the original bottle!), using clean spatulas each time

6. How to use a pipette correctly, including how to read and manipulate it

7. Making chemical solutions; provide guide sheets for:
   a. Solution preparation
   b. Molarity calculations
   c. Dilutions

8. Understanding the importance and practice of sterile technique

9. Media preparation and how to use an autoclave

10. Literature research skills

11. Basic microbiology including:
    a.   Plating for single colonies
    b.   Growing liquid cultures
    c.   Growth conditions for an organism

12. Basic molecular biology techniques including:
    a.   DNA isolation
    b.   Proper use of restriction enzymes

13. DNA isolation:
    a.   How to avoid contaminating DNA/RNA free/autoclaved materials
    b.   How to open microfuge tubes properly
    c.   How to label reagents

14. Basic guidelines for generating graphs and tables

*This list was compiled by mentors working primarily in molecular biology and microbiology labs. If your mentee's research does not use molecular biology techniques, you may wish to generate a list that is more relevant to your field.

Dear Mentor:

During the course of your discussions and writing assignments as mentors, The Wisconsin Program for Scientific Teaching has been compiling a series of mentoring case studies. The program would like to use these case studies for future mentoring discussion groups, and to supplement a book it has prepared on mentoring. In all instances, the author of each case study and any person(s) mentioned in it would remain anonymous. If your case study is chosen to be included, the program would ask for your approval of the text prior to distribution if you so desire.

Please sign the form below and return it to your discussion facilitator if you are willing to grant The Wisconsin Program for Scientific Teaching permission to use any of your case studies either verbally or in writing.

_____

I grant The Wisconsin Program for Scientific Teaching permission to use my case studies in future discussion groups and to supplement a book on mentoring, whether Web-based or in print. I understand that I will remain anonymous, as will any individuals mentioned in this work.

Would you like to be contacted for approval of any text based on your case study prior to distribution? _____ NO   _____ YES

EMAIL _____

Signature _____ *Date*_____

Print _____

EDUCATION

Scientific Teaching

Jo Handelsman,[1]* Diane Ebert-May,[2] Robert Beichner,[3] Peter Bruns,[4] Amy Chang,[5] Robert DeHaan,[6,†] Jim Gentile,[7] Sarah Lauffer,[1] James Stewart,[8] Shirley M. Tilghman,[9] William B. Wood[10]

Since publication of the AAAS 1989 report "Science for all Americans" (*1*), commissions, panels, and working groups have agreed that reform in science education should be founded on "scientific teaching," in which teaching is approached with the same rigor as science at its best (*2*). Scientific teaching involves active learning strategies to engage students in the process of science and teaching methods that have been systematically tested and shown to reach diverse students (*3*).

Given the widespread agreement, it may seem surprising that change has not progressed rapidly nor been driven by the research universities as a collective force. Instead, reform has been initiated by a few pioneers, while many other scientists have actively resisted changing their teaching. So why do outstanding scientists who demand rigorous proof for scientific assertions in their research continue to use and, indeed, defend on the basis of the intuition alone, teaching methods that are not the most effective? Many scientists are still unaware of the data and analyses that demonstrate the effectiveness of active learning techniques. Others may distrust the data because they see scientists who have flourished in the current educational system. Still others feel intimidated by the challenge of learning new teaching methods or may fear that identification as teachers will reduce their credibility as researchers (*3*).

This Policy Forum is needed because most scientists don't read reports but they do read *Science*. In addition, reports generally do not offer a guide to learning how to do scientific teaching, as we do with supporting online material (SOM) (*3*) and table (see page 522). We also present recommendations for moving the revolution forward.

Implementing Change in Lectures

Active participation in lectures and discovery-based laboratories helps students develop the habits of mind that drive science. However, most introductory courses rely on "transmission-of-information" lectures and "cookbook" laboratory exercises—techniques that are not highly effective in fostering conceptual understanding or scientific reasoning. There is mounting evidence that supplementing or replacing lectures with active learning strategies and engaging students in discovery and scientific process improves learning and knowledge retention (*3*).

Introductory classes often have high enrollments, frequently approaching 1000 students in biology courses. This need not be an impediment to scientific teaching. Many exercises that depart from traditional methods are now readily accessible on the Web, which makes it unnecessary for teachers to develop and test their own (*3*). Quantitative assessment indicates that these interactive approaches to lecturing significantly enhance learning, and although time allocated to inquiry-based activities reduces coverage of specific content, it does not reduce knowledge acquisition as measured by standardized exams (*4*).

Faculty are also using computer systems to engage students, assess learning, and shape teaching. Students can be asked to read and solve problems on a Web site, and their answers can be analyzed before class to guide the design of lectures (*3*).

Some scientists have replaced lectures almost entirely. Laws's course "Calculus-Based Physics Without Lectures" at Dickinson University (*5*) and Beichner's program, SCALE-UP, at North Carolina State University (see figure, this page) rely on a problem-based format in which students work collaboratively to make observations and to analyze experimental results. Students who learned physics in the SCALE-UP format at a

wide range of institutions demonstrated better problem-solving ability, conceptual understanding, and success in subsequent courses compared with students who had learned in traditional, passive formats (*3*).

These results are neither isolated nor discipline-specific. At the University of Oregon, Udovic showed dramatic differences between students taught biology in a traditional lecture and those taught "Workshop Biology," a series of active, inquiry-based learning modules (*6*). Similarly impressive results were achieved by Wright in a comparison of active and passive learning strategies in chemistry (*7*). Others have taught cross-disciplinary problem-based courses that integrate across scientific disciplines, such as Trempy's, "The World According to Microbes," at Oregon State University, which integrates science, math, and engineering. The course serves science ma-

A physics classroom at North Carolina State University arranged for traditional lectures (inset) and redesigned for group problem-solving in the SCALE-UP program.

jors and nonmajors, and outcome assessments indicate high content retention and student satisfaction (*8*).

Students as Scientists

Scientists of all disciplines have developed inquiry-based labs that require students to develop hypotheses, design and conduct experiments, collect and interpret data, and write about their results (*9*). Many of these involve simple, inexpensive materials configured so that they invite students to ask their own questions. In addition to labs that have already been tested in the classroom, resources are available to help teachers convert cookbook labs into open-ended, inquiry-based labs (*3*). Some schools provide introductory-level students with the opportunity to conduct original research in a professor's research lab rather than take a tradition-

[1]Howard Hughes Medical Institute Professor, Department of Plant Pathology, University of Wisconsin–Madison; [2]Department of Plant Biology, Michigan State University; [3]Department of Physics, North Carolina State University; [4]Howard Hughes Medical Institute; [5]American Society for Microbiology; [6]National Research Council; [7]Dean of Natural Sciences, Hope College; [8]Department of Curriculum and Instruction, University of Wisconsin–Madison; [9]President, Princeton University; [10]Department of Molecular Cellular, and Developmental Biology, University of Colorado at Boulder. For complete addresses, see SOM.

*To whom correspondence should be addressed. E-mail: joh@plantpath.wisc.edu.
†Present address: Division of Educational Studies, Emory University, Atlanta, GA, 30322, USA.

al classroom lab course (*3*). These opportunities are challenging for instructors, but teach students the essence of investigation.

How Universities Can Promote Change

Research universities should provide leadership in the reform movement. Faculty and administrators should collaborate to overcome the barriers and to create an educational ethos that enables change. We need to inform scientists about education research and the instructional resources available to them so that they can make informed choices. We must admit that citing our most successful students as evidence that our teaching methods are effective is simply not scientific. Instead, we need to apply innovative metrics to assess the outcomes of teaching. Controlled experiments and meta-analyses that compare student achievement with various teaching strategies provide a compelling basis for pedagogical choices (*10*), but the need for assessment extends into every classroom. Many tools to assess learning are available (*3*). Assessments of long-term retention of knowledge, entrance into graduate school, and employment and professional success should be included as well.

Research universities should overhaul introductory science courses for both science majors and nonmajors using the principles of scientific teaching. The vision should

SCIENTIFIC TEACHING EXAMPLES

Group problem-solving in lecture

www.ibscore.org/courses.htm

http://yucca.uoregon.edu/wb/index.html

http://mazur-www.harvard.edu/education/educationmenu.php

Problem-based learning

www.udel.edu/pbl/

www.microbelibrary.org

www.ncsu.edu/per/scaleup.html

http://webphysics.iupui.edu/jitt/jitt.html

Case studies

www.bioquest.org/lifelines/

http://ublib.buffalo.edu/libraries/projects/cases.case.html

http://brighamrad.harvard.edu/education/online/tcd/tcd.html

Inquiry-based labs

www.plantpath.wisc.edu/fac/joh/bbtl.htm

www.bioquest.org/

http://biology.dbs.umt.edu/biol101/default.htm

http://campus.murraystate.edu/academic/faculty/terry.derting/ccli/cclihomepage.html

Interactive computer learning

www.bioquest.org/

www.dnai.org

http://evangelion.mit.edu/802TEAL3D/

http://ctools.msu.edu/

originate from departments and be supported by deans and other academic administrators. Science departments should incorporate education about teaching and learning into graduate training programs and should integrate these initiatives into the educational environment and degree requirements. This could include, for example, development of peer-reviewed instructional materials based on the student's thesis research. Funding agencies have a responsibility to promote this strategy. National Institutes of Health and the National Science Foundation should, for example, require that graduate students supported on training grants acquire training in teaching methods, just as the NIH has required training in ethics.

Universities need to provide venues for experienced instructors to share best practices and effective teaching strategies. This will be facilitated, in part, by forming educational research groups within science departments. These groups might be nucleated by hiring tenure-track faculty who specialize in education, as 47 physics departments have done in the past 6 years. Other strategies include incorporating sessions about teaching into their seminar series, developing parallel series about teaching, or establishing instructional material "incubators" where researchers incorporate research results into teaching materials with guidance from experts in pedagogy. The incubators would provide an innovative mechanism to satisfy the "broader impact" mandate in research projects funded by the NSF.

Universities should place greater emphasis on awareness of new teaching methods, perhaps ear-marking a portion of research start-up packages to support attendance of incoming instructors at education workshops and meetings. Deans and department chairs at Michigan State University and University of Michigan have found that this strategy sends a message to all recruits that teaching is valued and it helps with recruiting faculty who are committed to teaching.

Distinguished researchers engaged in education reforms should exhort faculty, staff, and administrators to unite in education reform and should dispel the notion that excellence in teaching is incompatible with first-rate research. Federal and private funding agencies have contributed to this goal with programs such as the NSF's Distinguished Teaching Scholar Award and the Howard Hughes Medical Institute Professors Program, which demonstrate that esteemed researchers can also be innovative educators and bring prestige to teaching.

Universities and professional societies need to create more vehicles for educating faculty in effective teaching methods. For example, the National Academies Summer

Institutes on Undergraduate Education, the Council of Graduate Schools' Preparing Future Faculty program, the American Society for Microbiology Conference for Undergraduate Educators, and Workshops for New Physics and Astronomy Faculty are steps toward this goal (*3*).

Finally, the reward system must be aligned with the need for reform. Tenure, sabbaticals, awards, teaching responsibilities, and administrative support should be used to reinforce those who are teaching with tested and successful methods, learning new methods, or introducing and analyzing new assessment tools. This approach has succeeded at the University of Wisconsin–Madison, which has rewritten tenure guidelines to emphasize teaching, granted sabbaticals based on teaching goals, and required departments to distribute at least 20% of merit-based salary raises based on teaching contributions (*3*).

If research universities marshal their collective will to reform science education, the impact could be far-reaching. We will send nonscience majors into society knowing how to ask and answer scientific questions and be capable of confronting issues that require analytical and scientific thinking. Our introductory courses will encourage more students to become scientists. Our science majors will engage in the process of science throughout their college years and will retain and apply the facts and concepts needed to be practicing scientists. Our faculty will be experimentalists in their teaching, bringing the rigor of the research lab to their classrooms and developing as teachers throughout their careers. Classrooms will be redesigned to encourage dialogue among students, and they will be filled with collaborating students and teachers. Students will see the allure of science and feel the thrill of discovery, and a greater diversity of intellects will be attracted to careers in science. The benefits will be an invigorated research enterprise fueled by a scientifically literate society.

References and Notes

1. AAAS, "Science for all Americans: A Project 2061 report on literacy goals in science, mathematics, and technology" (AAAS, Washington, DC, 1989).
2. AAAS, "The Liberal Art of Science" (AAAS, Washington, DC, 1990).
3. Supporting online material provides further references on this point.
4. D. Ebert-May et al., Bioscience 47, 601 (1997).
5. P. Laws, Phys. Today 44, 24 (1991).
6. D. Udovic et al., Bioscience 52, 272 (2002).
7. J. C. Wright et al., J. Chem. Educ. 75, 986 (1998).
8. J. Trempy et al. Microbiol. Educ. 3, 26 (2002).
9. J. Handelsman et al., Biology Brought to Life: A Guide to Teaching Students How to Think Like Scientists (McGraw-Hill, New York, 1997).
10. L. Springer et al., Rev. Educ. Res. 69, 21 (1999).
11. We thank C. Matta, C. Pfund, C. Pribbenow, A. Fagen, and J. Labov for comments and A. Wolf for contributions to the supplemental materials. Supported in part by the Howard Hughes Medical Institute.

Supporting Online Material
www.sciencemag.org/cgi/content/full/304/5670/521/DC1

Teaching Scientists to Teach

We should train graduate students to be educators as well as researchers.

BY JO HANDELSMAN

Imagine if music schools trained pianists to play with only the right hand, leaving them on their own to figure out the left hand's responsibility. Ridiculous? Yes. But that is not unlike the way research universities train scientists.

On the one hand, so to speak, research-university graduates excel at doing science, given their institutions' focus on rigor, intensity and high standards in the practice of scientific research; on the other hand, they emerge largely untrained to teach science—to the public, to students generally and even to the next generation in their own fields—simply because graduate programs pay little attention to teaching scientists to teach.

The future scientist's teacher training, such as it is, is a casual and ad hoc affair with little design in the process or passion in the delivery. Some students serve as teaching assistants or mentors for undergraduates; others don't. Some receive supervision while engaged in teaching activities; others are left to learn—or flounder—on their own. It is unimaginable that students would complete the nation's best graduate science programs unable to deliver a compelling research seminar, defend an experimental design or write a scientific paper. Likewise, we ought to require that our graduate students also know how to craft a lecture, design a pedagogically sound learning exercise, successfully mentor an undergraduate student and communicate science to broad audiences.

In short, as we train the next generation of scientists, we should help students develop skills as educators—and expect that in that pursuit they would aspire to the same levels of knowledge, creativity and spirit of experimentation that we require of their research.

Whether they formally teach or not, scientists need to explain and make science compelling to nonscientists—industrial managers, government policymakers, patent examiners, the world. Every researcher has a responsibility to share his or her results with the public that supports the research and uses its products. With sound instruction in the art of teaching, scientists will be much better equipped to meet this responsibility. And those who enter the professoriate, where teaching is an explicit job requirement, will do so with skill and grace, having developed a theoretical framework about learning, cognition and the objectives of science education as well as a toolbox of teaching techniques to draw upon. Thus, strong teaching skills strengthen a Ph.D. scientist's career, whatever direction it may take.

SCAFFOLDING FOR GROWTH

Some might say there is no spare time in graduate education—for graduate students to master their discipline's rapidly expanding knowledge base is challenge enough. But training students to teach will not add years to their degree programs. Just a single semester of learning and practicing teaching as part of an intense, supportive and critical community, can build ample scaffolding for a student's future growth as a teacher. And for graduate students who are flagging or unfocused, successful teaching may renew a love of science. Their teaching can stimulate them to spend more time in the lab, plan their work with greater care and effectively direct the resources available, including the undergraduates they mentor.

Graduates of U.S. research universities become faculty at both undergraduate education institutions and research universities. Thus, if their own mentors embrace the goal of training graduate students in the art and science of teaching, the effect will cascade through the higher-education system. Such reform would improve the education of undergraduates at all institutions of higher learning, leading to a citizenry that not only has an enhanced sense of the power and limits of scientific inquiry but can also profit from the intellectual and experimental foundations of that inquiry. Programs by public and private agencies, including the HHMI Professors Program, help stimulate such important reforms.

We need to adjust our priorities and correct this historic imbalance of learning how to practice science but not how to teach it. In so doing, we will educate an entirely new generation of scientists who offer improved classroom teaching and more accessible public communication about science. That, in turn, will foster more informed discussion about the myriad science-rich issues that are unfolding before us at an ever-escalating pace, and wiser use of our country's resources, both material and human.

Research universities should raise a generation of future scientists who, like pianists who play with both hands, practice their art with a dynamic complement of skills, to the great benefit of society. ∎

Jo HANDELSMAN *is an HHMI Professor at the University of Wisconsin–Madison. Her research focuses on the structure, function and networks of microbial communities.*

Session 2:
Learning to Communicate

Discussion Outline: Session 2

Topics:

Case Studies: Projects

Mentoring Philosophies

Describe Assignments for Session 3: Establishing a
Relationship and Expectations

Materials for Mentors:

Case Studies: Projects

Mentor-Mentee Check-in Questions: Establishing a Relationship

Mentor-Mentee Check-in Questions: Defining Your Path

"What Mentors Do"

Session 2:

Discuss Case Study: Project Selection

When using either of these cases, present the mentors with two questions:

1. *If you were the undergraduate student, how would you feel?*
2. *If you were the faculty adviser, what would you do?*

Thoughts that have surfaced in previous discussions using the first case are:

Undergraduate Student Perceptions:
- Project choice showed favoritism
- Some projects are "cool," others are not
- Some projects are not important to the lab's larger goals
- Some projects are slower than others
- Mark's mentor may be better, so the project seems more appealing
- Other projects may be more collaborative, so they seem more appealing
- Overall, the student feels insulted and not respected

Advice to the Adviser:
- Be flexible
- Build a molecular element into the project
- Let the student "grow into" the challenge, i.e., if you get "x" to work, you can do "y"
- Let them try other techniques
- Improve communication with the student
- Deal with sulkiness early on

How do you feel about the project your mentee will be working on?

Some leading questions might be:

1. How do you feel about the project you have given to your mentee?

2. How do you think your mentee feels about the project?

3. Do you feel the project is a "good project," given the parameters we identified during our previous session?

4. Does anyone have comments about the projects of other members of the group?

5. In light of the above case study, what will you do if your student does not like their project?

6. What can you do if a student develops a new project idea?

Discussion of Mentoring Philosophies

Some guiding questions we have used to facilitate an open discussion on this topic are:

1. What is a mentor?

2. What are some common themes among the philosophies?

3. What is the difference between a teacher and a mentor?

4. What kinds of mentors are there?

5. What kind of mentor do you want to be?

6. Can a mentor also be an evaluator? Are there conflicting power issues in this relationship?

7. What kind of "power" does a mentor have?

What do you expect from your mentee and what do they expect from you?

The mentor and the mentee need to establish clear expectations in the beginning of the relationship and to revisit the discussion of expectations often.

One leading question that has proven useful is, "What do you expect from your mentee and what do they expect from you?" Asking this question of the group and compiling a list of expectations may help mentors appreciate the wide variety of expectations they may have. These expectations range from expecting a student to be punctual to expecting that a student will complete a certain experiment.

Assignments

1. Ask each mentor to interview their mentee and write a brief biography. This assignment is effective in helping to establish a connection between the mentor and mentee beyond the research project. Some guiding questions for this assignment can be found in this section.

2. Encourage the mentors and mentees to share their expectations with one another. Specific guiding questions to help the mentor and mentee in this discussion can be found in this section.

3. (optional) Consider asking mentors to have their mentees write letters of recommendation for themselves, including the items they hope their mentor will be able to address at the conclusion of the research experience.

Case Study: Projects

" I mentored an undergraduate student who came from another university for the summer. I explained the project to him and taught him how to make media and grow bacteria. Because my professor and I did not think he had sufficient genetics background for a molecular project, we gave him a microbiology project.

He was very quiet for the first ten days of the project and then he went to my adviser and complained about the project. He said he wanted a project "like Mark's." Mark was a student with a strong genetics background and his project was to clone and sequence a gene. My adviser insisted that my mentee keep the project I had designed for him, but the student became sulky. As the summer went on and he didn't get any of his experiments to work, I began to wonder if he understood what we were doing or even cared about it. "

Case Study: Projects

❝ I am a graduate student in a large lab. A week ago, an undergraduate student joined me to do an independent summer research project. He really wanted to come to our lab and aggressively sought us out, which I assumed was because of our field of research. He had seen presentations about our lab's research and had read some of our major papers, so he knew what we worked on. This young man was clearly intelligent, and he knew what he wanted out of a research experience. He was exactly the type of student I would love to see go to graduate school. Moreover, he was a first-generation college student.

My adviser and I came up with two aspects of my research compatible with the undergraduate's interests that would be feasible for him to work on in the short, eight-week summer session. When he arrived, I presented the two ideas to him, gave him several papers to read, and told him to let me know by the end of the week which project he preferred. He seemed lukewarm about both projects and, when he returned the next day, he enthusiastically presented his idea for a different project. It was related to what we do, but branched into a field that my adviser was not funded for and about which I knew little. I didn't want to squash his enthusiasm, and wanted to reinforce his creativity and independence, but I felt overwhelmed by the prospect of learning an entirely new field in order to advise him well. Moreover, my adviser was concerned that the agency that funds our work would likely not be supportive of this new area from another lab. With only seven weeks of the summer research program remaining before his poster presentation, I was stumped. ❞

Establishing a Relationship

Goals:

- Get to know one another.

- Begin to define your working relationship and establish expectations.

- Define the goals of your summer research project.

Students (Mentees):

- Who are you? Where is your home? How/when did you become interested in a career in science?

- What is your major and what are your future career plans?

- Why do you want to do research and how will it help you reach your career goals?

- What would success in this research program look like to you?

- Do you have any previous research experience? If so, what did you do? What did you like about it? What did you dislike about it?

  How do you learn best (e.g., hands-on experience, reading literature about a topic, verbal explanations, process diagrams, etc.)? What is the most useful kind of assistance your mentor can provide?

- Do you prefer to work alone or in groups? What kind of group or collaborative work experience have you had?

- Do you have any questions about the background reading your mentor sent you before the start of the program?

Mentors:

- Who are you? How did you become a scientist?

- Why have you chosen to be an undergraduate research mentor? What do you hope to gain from this experience?

- What would success in this research program look like to you? What skills (technical, communication) should your mentee develop?

- Who are the people who work in your lab? What are their responsibilities and how should your mentee expect to interact with each of them? What are the proper channels of communication?

- How many hours per week do you expect your mentee to work in the lab? Are there specific times of day that you expect your student to be in the lab?

- What is your teaching style? How do you prefer to help students learn to conduct research? Is there a process that you normally follow?

These guidelines were developed by Janet Branchaw, Center for Biology Education, University of Wisconsin, based on Zachary, L.J. (2000). *The Mentor's Guide: Facilitating Effective Learning Relationships*. San Francisco, CA: Jossey-Bass, Inc., Publishers.

Defining Your Path

Goals:

- Reaffirm expectations between mentor and student.

- Clearly define the research project and a timeline for completion of specific experiments.

Students (Mentees):

- What do you like best about working in your lab so far?

- What do you find most challenging about working in your lab? How can your mentor help you deal with this?

- What have you learned about working in a lab that you did not expect before arriving on campus?

- Are you comfortable working with the other members of your laboratory? If not, how can your mentor facilitate these interactions?

- Now that you have thought about your research proposal, what aspects of the research project are still unclear to you? What aspects are the most exciting and interesting?

- Which of the research techniques that you will learn, or have learned, do you find most challenging?
  How can your mentor facilitate your learning this technique?

- How much time do you expect it will take to complete your research project?

- Would you like to be able to spend more time with your mentor? Do you feel you are ready to work more independently?

Mentors:

- What do you see as your mentee's greatest strength(s) in the laboratory so far?

- What area(s) do you think your mentee should focus on developing? How do you suggest they do this, and how can you facilitate this process?

- How much time do you expect it will take to complete your mentee's research project?

- What have you learned about working with your mentee that you did not expect to learn?

These guidelines were developed by Janet Branchaw, Center for Biology Education, University of Wisconsin, based on Zachary, L.J. (2000). *The Mentor's Guide: Facilitating Effective Learning Relationships.* San Francisco, CA: Jossey-Bass, Inc., Publishers.

Session 3:
Goals & Expectation

Discussion Outline: Session 3

Topics:

Case Study: Independence

Case Studies: Trust

Discussion Questions:

- Are you and your mentee clear on expectations?

- How do you know they understand what you are saying?

Materials for Mentors:

Case Study: Independence

Case Studies: Trust

Case Study: Respect and Trust

Case Study: Ethics

Reading: "What is a Mentor?"

Session 3:

Discussion of Expectations

Discuss the topic of expectations and hear how each mentor's discussion of expectations went with their mentee. Remember that some guiding questions on expectations can be found in the previous section. In addition, a case involving independence is included, and may be used to help mentors recognize the importance of fostering independence in their mentees.

Discussion of Case Study: Trust

When using the first case in this section focusing on trust, present mentors with the following question:

If you were the graduate student mentor, how would you feel?

Thoughts that have surfaced in previous discussions are:

- Adviser has undermined the mentor's authority

- Mentor will not confide in adviser again

- Adviser has undermined the undergraduate's confidence

- The undergraduate is now labeled as a slob and this may prevent a change in behavior.

Guiding questions:
- Should the mentor have approached their adviser with this issue?

- What should the graduate student do to alter the outcome?

- If you were the adviser, how would you have handled the situation?

- How does this type of situation affect the lab environment?

Other cases involving similar topics can be found in this section.

How do you know if they understand what you are saying?

To facilitate this discussion, we have asked the mentors to suggest strategies and generate a list. Some strategies mentors have suggested are:

- Have them explain their project back to you.

- Have them explain their project to another undergraduate in the lab.

- Have them draw a flowchart or diagram of the project or write a paragraph describing the project.

- Ask another member of the lab to ask the student to explain the project.

- Develop some work sheets for them to complete that assess understanding; work sheets can also be given to accompany scientific papers you ask the students to read.

- If a student makes an assertion in their explanation, have them search the literature to verify it.

Case Study: Independence

66 An experienced undergraduate researcher was constantly seeking input from the mentor on minor details regarding his project. Though he had regular meetings scheduled with the mentor, he would bombard her with several e-mails daily or seek her out anytime she was around, even if it meant interrupting her work or a meeting that was in progress. It was often the case that he was revisiting topics that had already been discussed. This was becoming increasingly frustrating for the mentor, since she knew the student was capable of independent work (having demonstrated this during times she was less available). The mentor vented her frustration to at least one other lab member and wondered what to do. **99**

Session 3

Case Study: Trust 1

" A graduate student mentor was frustrated because her student was not running successful experiments. While the undergraduate had great enthusiasm for the project, each experiment failed because of some sloppy error—forgetting to pH the gel buffer, forgetting to add a reagent to a reaction, or forgetting to turn down the voltage on a gel box.

After a month of discussions, and careful attempts to teach the student habits that would compensate for his forgetfulness, the graduate student was ready to give up. She spoke with her adviser and asked for advice, hoping that she could fix the problem and start getting useful data from her undergraduate. The adviser offered to work with the undergraduate mentee. When the undergraduate walked into his office, the faculty member said, "I hear you're a slob in the lab. You gotta clean up your act if we're going to get any data out of you." Seeing the crushed and humiliated look on the undergraduate's face, he quickly added, "I'm a slob too—that's why I'm in here pushing papers around and not in the lab doing the hard stuff like you guys!" "

Case Study: Trust 2

" As a graduate student, I supervised an undergraduate in a summer research program. At the end of the summer, my adviser said we should publish a paper that included some of the work done by the undergraduate. I got nervous because I thought I could trust the undergraduate, but I wasn't totally sure. He seemed very eager to get a particular answer and I worried that he might have somehow biased his collection of data. I didn't think he was dishonest, just overeager. My question is: should I repeat all of the student's experiments before we publish? Ultimately, where do we draw the line between being trusting and not knowing what goes into papers with our names on them? "

Case Study: Trust & Respect

66 My adviser accepted a student for an undergraduate research experience without asking any of the graduate students if we had time for her. She was assigned to the most senior graduate student for mentoring, but he was in the process of writing his dissertation and had no time to help her with a project. He asked me if I would take her on and have her help me with my research project. I agreed, assuming that I was now her mentor and not understanding that she was expected to produce a paper and give a presentation on her research at the end of the summer.

We worked together well initially as I explained what I was doing and gave her tasks that taught her the techniques. She didn't ask many questions, nodded when I asked if she understood, and gave fairly astute answers when asked to explain the reason for a particular method.

I became frustrated as the summer progressed, though. Instead of asking me questions, she went to the senior graduate student for help on my project. He did not know exactly what I was doing, but didn't let me know when he and she were meeting. He even took her in to our adviser to discuss the project, but didn't ask me to be involved. As more of this occurred, the student became quieter around me, didn't want to share what she had done while I was out of the lab, and acted as though there was a competition with me for obtaining the sequence, rather than it being a collaborative effort. I didn't think too much about this and didn't recognize the conflict. She obviously didn't like sharing the project with me, which was even more evident when she wrote the paper about our research without including my name. She didn't want to give me a copy of the draft to review and I only obtained a copy by cornering the senior graduate student after I overheard them discussing the methods section and asked for a copy. I wasn't provided a final version of the paper nor was I informed of when or where she was presenting the research until two days before her presentation when I happened to see her practicing it with the senior student.

I felt very used throughout the process and disappointed that I didn't see what was occurring and address it sooner. In fact, I am not sure if addressing it would have solved the problems I had—being stuck in between a student and the person she saw as her mentor. The difficult thing, for me at least, is identifying that there is a problem before it is too late to bow out or to bring all parties to the table to discuss a different approach to the mentoring. Do you have any suggestions for me? I don't ever want to encounter this again and would like to head it off as soon as I can recognize that it is occurring. **"**

Case Study: Ethics

Your mentee, James, is a high school student who has grand aspirations of one day becoming a doctor. He has participated in science fair opportunities since the seventh grade. He has taken the advice of educational professionals to gain lab experience in order to make his college entrance application look distinguished. He worked with you this past summer and recently has asked if he can do a science fair project in your lab. You are asked to sign the abstract of the project. Because of divergent school and project deadlines, the abstract is due before the experiment is completed.

One month prior to the fair, you notice that he has not really been in the lab doing the work. When you question him, he is vague about what he is doing. It is unclear that he is doing anything at all. On the day of the fair, you are surprised to see him there. His project's results win him a first-place award, giving him the opportunity to go to the state competition. You have the uncomfortable feeling that he has not done the work.

How do you feel toward this student?

What would/could you do next?

How quickly do you have to act?

When is it not a good time to act?

What are your objectives and goals in this situation?

A few days later, you ask to meet with James and his teacher (explaining to the teacher your reservations, but still making no accusations). At that interview, James is very uncomfortable, but rather vaguely answers all of your questions. He brings his overheads from the presentation to that meeting for review, but he does not bring his notebook (which is technically property of the lab). You leave that meeting with stronger suspicions, but no proof. You request that he return his notebook to the lab. He signs a statement that the results of the project were his work and reported accurately.

What would/could you do next?

How much time can you/should you legitimately spend on this matter?

What are legitimate actions you can take when you have unsubstantiated suspicions? Is it OK to act on them? Why or why not?

How do you combat the thought: "but I know lots of others who do the same thing, or have done worse?"

Through James' teacher, you request the notebook and results again in order to "confirm" his results before they are presented at the statewide competition. Two days later, James comes into your office, and nervously asks to talk to you about the project. He says there was a lot of pressure on him, and he ran out of time, and he is ashamed, a but he "twisted" the data. He apologizes, says his teacher is withdrawing his first-place award, and he wants to redeem himself in some way; he knows what he did is wrong.

How do you feel toward this student?

What would/could you do next?

How quickly do you have to act?

When is it not a good time to act?

What are your objectives and goals in this situation?

ADVISER, TEACHER, ROLE MODEL, FRIEND

On Being a Mentor to Students in Science and Engineering

Session 3

NATIONAL ACADEMY OF SCIENCES

NATIONAL ACADEMY OF ENGINEERING

INSTITUTE OF MEDICINE

WHAT IS A MENTOR?

The notion of mentoring is ancient. The original Mentor was described by Homer as the "wise and trusted counselor" whom Odysseus left in charge of his household during his travels. Athena, in the guise of Mentor, became the guardian and teacher of Odysseus' son Telemachus.

In modern times, the concept of mentoring has found application in virtually every forum of learning. In academics, *mentor* is often used synonymously with *faculty adviser*. A fundamental difference between mentoring and advising is more than advising; mentoring is a personal, as well as, professional relationship. An adviser might or might not be a mentor, depending on the quality of the relationship. A mentoring relationship develops over an extended period, during which a student's needs and the nature of the relationship tend to change. A mentor will try to be aware of these changes and vary the degree and type of attention, help, advice, information, and encouragement that he or she provides.

In the broad sense intended here, a mentor is someone who takes a special interest in helping another person develop into a successful professional. Some students, particularly those working in large laboratories and institutions, find it difficult to develop a close relationship with their faculty adviser or laboratory director. They might have to find their mentor elsewhere—perhaps a fellow student, another faculty member, a wise friend, or another person with experience who offers continuing guidance and support.

In the realm of science and engineering, we might say that a good mentor seeks to help a student optimize an educational experience, to assist the student's socialization into a disciplinary culture, and to help the student find suitable employment. These obligations can extend well beyond formal schooling and continue into or through the student's career.

The Council of Graduate Schools (1995) cites Morris Zelditch's useful summary of a mentor's multiple roles: "Mentors are advisers, people with career experience willing to share their knowledge; supporters, people who give emotional and moral encouragement; tutors, people who give specific feedback on one's performance; masters, in the sense of employers to whom one is apprenticed; sponsors, sources of information about and aid in obtaining opportunities; models, of identity, of the kind of person one should be to be an academic."

In general, an effective mentoring relationship is characterized by mutual respect, trust, understanding, and empathy. Good mentors are able to share life experiences and wisdom, as well as technical expertise. They are *good listeners, good observers,* and *good problem-solvers.* They make an effort to know, accept, and respect the goals and interests of a student. In the end, they establish an environment in which the student's accomplishment is limited only by the extent of his or her talent.

The Mentoring Relationship

The nature of a mentoring relationship varies with the level and activities of both student and mentor. In general, however, each relationship must be based on a common goal: to advance the educational and personal growth of the student. You as mentor can also benefit enormously.

There is no single formula for good mentoring; mentoring styles and activities are as varied as human relationships. Different students will require different amounts and kinds of attention, advice, information, and encouragement. Some students will feel comfortable approaching their mentors; others will be shy, intimi-

---

WHY BE A GOOD MENTOR?

The primary motivation to be a mentor was well understood by Homer: the natural human desire to share knowledge and experience. Some other reasons for being a good mentor:

Achieve satisfaction. For some mentors, having a student succeed and eventually become a friend and colleague is their greatest joy.

Attract good students. The best mentors are most likely to be able to recruit—and keep—students of high caliber who can help produce better research, papers, and grant proposals.

Stay on top of your field. There is no better way to keep sharp professionally than to coach junior colleagues.

Develop your professional network. In making contacts for students, you strengthen your own contacts and make new ones.

Extend your contribution. The results of good mentoring live after you, as former students continue to contribute even after you have retired.

dated, or reluctant to seek help. A good mentor is approachable and available.

Often students will not know what questions to ask, what information they need, or what their options are (especially when applying to graduate programs). A good mentor can lessen such confusion by getting to know students and being familiar with the kinds of suggestions and information that can be useful.

In long-term relationships, friendships form naturally; students can gradually become colleagues. At the same time, strive as a mentor to be aware of the distinction between friendship and favoritism. You might need to remind a student—and yourself—that you need a degree of objectivity in giving fair grades and evaluations. If you are unsure whether a relationship is "too personal," you are probably not alone. Consult with the department chair, your own mentor, or others you trust. You might have to increase the mentor-student distance.

Students, for their part, need to understand the professional pressures and time constraints faced by their mentors and not view them as merely a means—or impediment—to their goal. For many faculty, mentoring is not their primary responsibility; in fact, time spent with students can be time taken from their own research. Students are obliged to recognize the multiple demands on a mentor's time.

At the same time, effective mentoring need not always require large amounts of time. An experienced, perceptive mentor can provide great help in just a few minutes by making the right suggestion or asking the right question. This section seeks to describe the mentoring relationship by listing several aspects of good mentoring practice.

Careful listening. A good mentor is a good listener. Hear exactly what the student is trying to tell you—without first interpreting or judging. Pay attention to the "subtext" and undertones of the student's words, including tone, attitude, and body language. When you think you have understood a point, it might be helpful to repeat it to the student and ask whether you have understood correctly. Through careful listening, you convey your empathy for the student and your understanding of a student's challenges. When a student feels this empathy, the way is open for clear communication and more-effective mentoring.

Keeping in touch. The amount of attention that a mentor gives will vary widely. A student who is doing well might require only "check-ins" or brief meetings. Another student might have continuing difficulties and require several formal meetings a week; one or two students might occupy most of an adviser's mentoring time. Try through regular contact—daily, if possible—to keep all your students on the "radar screen" to anticipate problems before they become serious. Don't assume that the only students who need help are those who ask for it. Even a student who is doing well could need an occasional, serious conversation. One way to increase your

---

GOOD MENTORING: SEEKING HELP

A white male professor is approached by a black female undergraduate about working in his lab. She is highly motivated, but she worries about academic weaknesses, tells him she is the first member of her family to attend college, and asks for his help. He introduces her to black male colleague and a white female graduate student in related fields who offer to supplement his advice on course work, planning, and study habits. He also seeks information about fellowships and training programs and forwards this information to the student.

Comment: This student already has an essential quality for academic success—motivation. By taking a few well-planned steps, an alert mentor can help a motivated student initiate a network of contacts, build self-esteem, and fill academic gaps.

---

awareness of important student issues and develop rapport is to work with student organizations and initiatives. This will also increase your accessibility to students.

Multiple mentors. No mentor can know everything a given student might need to learn in order to succeed. *Everyone benefits from multiple mentors* of diverse talents, ages, and personalities. No one benefits when a mentor is too "possessive" of a student.

Sometimes a mentoring team works best. For example, if you are a faculty member advising a physics student who would like to work in the private sector, you might encourage him or her to find mentors in industry as well. A non-Hispanic faculty member advising a Hispanic student might form an advising team that includes a Hispanic faculty member in a related discipline. Other appropriate mentors could include other students, more-advanced postdoctoral associates, and other faculty in the same or other fields. A good place to find additional mentors is in the disciplinary societies, where students can meet scientists, engineers, and students from their own or other institutions at different stages of development.

Coordinate activities with other mentors. For example, a group of mentors might be able to hire an outside speaker or consultant whom you could not afford on your own.

Building networks. You can be a powerful ally for students by helping them build their network of contacts and potential mentors. Advise them to begin with you, other faculty acquaintances, and off-campus people met through jobs, internships, or chapter meetings of professional societies. Building a professional network is a lifelong process that can be crucial in finding a satisfying position and career.

Professional Ethics

Be alert for ways to illustrate ethical issues and choices. The earlier that students are exposed to the notion of scientific integrity, the better prepared they will be to deal with ethical questions that arise in their own work.

ADVICE FOR NEW MENTORS

For most people, good mentoring, like good teaching, is a skill that is developed over time. Here are a few tips for beginners:

☛ **Listen patiently.** Give the student time to get to issues they find sensitive or embarrassing.

☛ **Build a relationship.** Simple joint activities—walks across campus, informal conversations over coffee, attending a lecture together—will help to develop rapport. Take cues from the student as to how close they wish this relationship to be. (See "Sexual harassment" in section on Population-diversity issues.)

☛ **Don't abuse your authority.** Don't ask students to do personal work, such as mowing lawn, baby-sitting, and typing.

☛ **Nurture self-sufficiency.** Your goal is not to "clone" yourself but to encourage confidence and independent thinking.

☛ **Establish "protected time" together.** Try to minimize interruptions by telephone calls or visitors.

☛ **Share yourself.** Invite students to see what you do, both on and off the job. Tell of your own successes and failures. Let the student see your human side and encourage the student to reciprocate.

☛ **Provide introductions.** Help the student develop a professional network and build a community of mentors.

☛ **Be constructive.** Critical feedback is essential to spur improvement, but do it kindly and temper criticism with praise when deserved.

☛ **Don't be overbearing.** Avoid dictating choices or controlling a student's behavior.

☛ **Find your own mentors.** New advisers, like new students, benefit from guidance by those with more experience.

Discuss your policies on grades, conflicts of interest, authorship credits, and who goes to meetings. Use real-life questions to help the student understand what is meant by scientific misconduct: What would you do if I asked you to cut corners in your work? What would you do if you had a boss who was unethical?

Most of all, *show by your own example what you mean by ethical conduct.* You might find useful the COSEPUP publication *On Being a Scientist: Responsible Conduct in Research* (1995), also available on line.

Population-Diversity Issues

In years to come, female students and students of minority groups might make up the majority of the population from which scientists and engineers will emerge. Every mentor is challenged to adapt to the growing sex, ethnic, and cultural diversity of both student and faculty populations.

Minority issues. Blacks, Hispanics, and American Indians as a group make up about 23% of the US population, but only about 6% of the science and engineering labor force. Many minority-group students are deterred from careers in science and engineering by inadequate preparation, a scarcity of role models, low expectations on the part of others, and unfamiliarity with the culture and idioms of science. Mentors can often be effective through a style that not only welcomes, nurtures, and encourages questions, but also challenges students to develop critical thinking, self-discipline, and good study habits. Expectations for minority-group students in science have traditionally been too low, and this can have an adverse effect on achievement. A clear statement that you

---

POOR MENTORING: CULTURAL BIAS (1)

A foreign-born engineering student is reluctant to question his adviser. As a result, the adviser thinks the student lacks a grasp of engineering. The adviser tries to draw out the student through persistent questioning, which the student finds humiliating. Only the student's determination to succeed prevents him from quitting the program.

Comment: The student grew up in a country where he learned not to question or disagree with a person in authority. Had the adviser suspected that a cultural difference was at the root of the problem, he might have learned quickly why the student was reluctant to question him. When communication is poor, try to share yourself, listen patiently, and ask the students themselves for help.

POOR MENTORING: INAPPROPRIATE BEHAVIOR (2)

The male adviser of a female graduate student has not seen her for several months. Passing her in the hall, he squeezes her shoulder as he describes concerns about her research. He sends her an e-mail message, inviting her to discuss the project over dinner. She declines the invitation. He learns that she has redirected her work in a way he does not approve of, and he asks her to return to her original plan. He is astonished when she accuses him of sexual harassment and files a complaint with the dean's office.

Comment: In this case, the adviser erred in touching the student and extending a dinner invitation that could easily be misconstrued.

expect the same high performance from all students might prove helpful. Be aware of minority support groups on your campus and of appropriate role models. Link minority-group students with such national support organizations as the National Action Council for Minorities in Engineering (see "*Resources*").

Cultural issues. You could find yourself advising students of different cultural backgrounds (including those with disabilities) who have different communication and learning styles. Such students might hail from discrete rural or urban cultures in the United States or from abroad; in many programs, foreign-born students are in the majority. If you are not familiar with a particular culture, it is of great importance to demonstrate your willingness to communicate with and to understand each student as a unique individual. Are you baffled by a student's behavior? Remember that a cultural difference could be the reason. Don't hesitate to ask colleagues and the students themselves for help. Finding role models is especially important for students from a culture other than yours. Examine yourself for cultural biases or stereotypical thinking.

Female representation. In some fields—notably psychology, the social sciences, and the life sciences—females are well represented as students but underrepresented in the professoriate and are not always appointed to assistant professor positions at a rate that one would expect on the basis of PhD and postdoctoral student representation. In other fields—such as mathematics, physics, computer science, and engineering—females are underrepresented at all levels. In all fields, the confidence of female students might be low, especially where they are isolated and have few female role parent, suffering marital problems, or juggling the challenges of a two-career family. You might want to send a student to a colleague or counselor with special competence in family issues.

Sexual harassment. If you mentor a student of the opposite sex, extra sensitivity is required to avoid the appearance of sexual harassment. Inappropriate closeness between mentors and students will produce personal, ethical, and legal consequences not only for the persons involved but also for the programs or institutions of which they are part.

Be guided by common sense and a knowledge of your own circumstances. Is it appropriate to invite the student to discussions at your home? During meetings, should you keep the office door closed (for privacy) or open (to avoid the appearance of intimacy)? Make an effort to forestall misunderstandings by practicing clear communication. If you do have a close friendship with a student, special restrictions or self-imposed behavior changes might be called for.

But do not restrict students' opportunities to interact with you because of sex differences. In a respectful relationship, mutual affection can be an appropriate response to shared inquiry and can enhance the learning process; this kind of affection, however, is neither exclusive nor romantic. For additional guidance, talk with your department chair, your own mentor, or other faculty.

Disability issues. Students with physical, mental, emotional, or learning disabilities constitute about 9% of first-year students with planned majors in science and engineering. Be careful not to underestimate the potential of a student who has a disability. Persons with disabilities who enter the science and engineering workforce perform the same kinds of jobs, in the same fields, as others in the workforce. You should also keep in mind that persons with disabilities might have their own cultural background based on their particular disability, which cuts across ethnic lines.

As a mentor, you might be unsure how to help a student with a disability. Persons with disabilities can function at the same level as other students, but they might need assistance to do so. You can play a pivotal role in finding that assistance, assuring students that they are entitled to the assistance, and confirming they are able to secure assistance. Another very important role of the mentor is in making colleagues comfortable with students who have disabilities.

Many campuses offer programs and aids such as special counseling, special equipment (adaptive computer hardware, talking calculators, and communication devices), adapted physical education, learning disability programs, and academic support.

Further, your institution's specialist in Americans with Disabilities Act (ADA) issues might provide help (for example, in securing funding from the National Institutes of Health [NIH], the National Science Foundation [NSF], and other sources). However, keep in mind that this person might know less than you do about the needs of a student in your field—for example, in the use of particular equipment.

Remember that the student who lives with the disability is the expert and that you can ask this expert for help.

SUMMARY POINTS

☞ In a broad sense, a mentor is someone who takes a special interest in helping another develop into a successful professional.

☞ In science and engineering, a good mentor seeks to help a student optimize an educational experience, to assist the student's socialization into disciplinary culture, and to aid the student in finding suitable employment.

☞ A fundamental difference between a mentor and an adviser is that mentoring is more than advising; mentoring is a personal as well as a professional relationship. An adviser might or might not be a mentor, depending on the quality of the relationship.

☞ An effective mentoring relationship is characterized by mutual trust, understanding, and empathy.

☞ The goal of a mentoring relationship is to advance the educational and personal growth of students.

☞ A good mentor is a good listener.

☞ Everyone benefits from having multiple mentors of diverse talents, ages, and personalities.

☞ A successful mentor is prepared to deal with population-diversity issues, including those peculiar to ethnicity, culture, sex, and disability.

Session 4:
Identifying Challenges & Issues

Session 4

Discussion Outline: Session 4

Topics:

Discussion Questions

- Describe issues or challenges you are facing with your mentees.

- How do you know if there are problems?

Describe Assignment for Session 5: Proposed Solutions to Mentoring

Materials for Mentors:

Reading: "Mentoring: Learned, Not Taught"

Session 4:

Mentoring issues and challenges from the first few weeks

In this session, ask mentors to share mentoring challenges from the first few weeks. Discussion questions might include:

1. *What is the biggest challenge you are facing as a mentor?*

2. *What has been your biggest success as a mentor thus far?*

3. *What has been your biggest disappointment as a mentor thus far?*

Mentoring sessions in which mentors share their frustrations, chalenges, and successes have been the most enlightening. Try to give everyone a chance to talk. Often there is only time for one or two people to describe their mentoring case, but it is important that everyone has a voice in responding to these cases.

Alternately, try presenting the group with a challenge you are facing as a mentor. Ask them to help you decide how to handle the challenge.

How do you know if there are problems?

- Ask for honest feedback (see assignment below).

- Do not assume things are fine just because your mentee has not complained.

- When your mentee tells you things are fine, you may want to ask them to expand on that answer—i.e., ask them what they mean by "fine."

- Ask your mentee specifically what is going well and what is not going well.

Assignment

1. Ask mentors to choose one of the challenges they heard about in the session and propose one possible solution to share with the group.

2. Ask the mentors to discuss with their mentees the quality of their mentoring. Encourage the mentors to ask for honest feedback from their mentees.

Mentoring: Learned, Not Taught

Identifying Challenges

•

Jo Handelsman

Becoming a good mentor takes practice and reflection. Each of us tends to focus on certain aspects of mentoring, which we choose for many different reasons. Sometimes we focus on issues that were important to us as mentees, those we think are hard or uncomfortable to deal with (making us worry) or easy to handle (consequently making us feel good about our mentoring), or areas in which a mentee needs help. But few of us think about the diversity of issues that comprise the full mentoring experience, at least not when we are just starting out as mentors. By broadening our approach, and looking at mentoring in a systematic way, we can become more effective mentors more quickly than if we just confront the challenges as we stumble upon them. Some of us take decades to recognize all these facets of mentoring; others of us would never discover them on our own.

This chapter focuses largely on mentors of undergraduates and graduate students in a research lab, but many of the same issues arise in mentoring colleagues and others outside the lab. Each of us is likely to engage in numerous relationships as mentors and mentees throughout our careers and each relationship will be enhanced by what we learned in the last one. Reflecting on the following areas as your mentoring relationships evolve may help you avoid some common mistakes and hasten your arrival at a mentoring style and philosophy that is your own.

Mentoring principles, not practices, are universal

Although no one can provide formulas, practices, or behaviors that will work in every mentoring situation, there are some principles that should always guide mentoring relationships. It's a good idea to ask yourself periodically whether you are adhering to the basic principles you believe in. The values that most scientists would agree are inviolate in any mentoring relationship are: honesty, kindness, caring, and maintenance of high ethical and scientific standards. As you consider the differences among students, and design your mentoring strategies to serve them best, examine your values.

Mentees are different...from each other and from us

The diversity that our students bring us sustains the vibrancy of the scientific community and of science itself. Although most of us believe this in the abstract, dealing with people who are different from us or from our mental image of the ideal student can be frustrating and baffling. Those of us who are very organized, punctual, polite, tidy, diligent, smart, socially adept, witty, verbal, creative, confident, and tenacious probably value those characteristics in ourselves. When confronted with a mentee lacking any of them, we may wonder if they are cut out to be a scientist. Moreover, cognitive styles (the ways that we learn or think about problems) are often what scientists value most highly in themselves, but cognitive styles are idiosyncratic; thus, being a good mentor necessitates accommodating a style that differs from our own.

After we have worked with a student for a few weeks or months, we may begin to see performance issues that didn't emerge immediately. Some issues are small, some global. We may find that it drives us nuts that a student likes to work from noon until midnight, whereas we prefer working in the early morning. Or a student may seem unable to articulate the objectives of a research project even after substantial discussion and reading. Or the student may seem unable to get a product from PCR. Or come up with an idea of their own. There are no simple prescriptions for what to do. The following sections offer some questions for reflection and sample situations to provoke thought about dealing with these very complex, very human mentoring challenges.

Session 4

Building confidence

Probably the most important element of mentoring is learning that performance is the product of a complex interaction among innate ability, experience, confidence, education, and the nature of the performance environment. We have all had the experience of saying something eloquently and smoothly in one setting and then stuttering our way through the same words in a stressful setting. We have the ability to formulate the idea and express it well, but the stressful situation affects our performance. This happens to people in so many ways. If we are told as children that we are very smart, we develop confidence in our intelligence. In contrast, if we are told that we can't do science because we are female or a member of the wrong ethnic group, we may have lingering doubts even when we reach the highest levels of achievement. If we come from a family in which we are the first to go to college, we may feel that we just don't quite fit in when we are in the academic environment. All of these insecurities will surface at the most stressful times—when

> " ... performance is the product of a complex interaction among innate ability, experience, confidence, education, and the nature of the performance environment ... "

things aren't going well in the lab, when we are getting ready for exams, when we receive a poor grade, when our grants aren't funded and our papers are rejected. Those are the times when a mentor can make a difference. People with stores of confidence fall back on internal reinforcement during the rough times. The voice of a parent or teacher from the past saying "you can do it" may get them through. But people who haven't received those messages may need to hear them from a trusted mentor or colleague in order to keep going.

The challenge for many of us is not to fall into the habit of measuring every student against our own strengths. Most of us have the impulse to think, "I never needed so much support or coddling, why should I have to give it to my students?" or "Can they really make it in science with such a need for reinforcement and coaching?" But the job of a mentor is to set high standards for mentees and then help them meet those standards. One of the most satisfying parts of mentoring is the frequency with which students surprise us. So often we hear a colleague say that, although they pushed a student to be great, it was a surprise when the student actually became great. A mentor may help a student develop the skills to be an outstanding scientist, but the most important message a mentor can ever send is that they have faith that the mentee will succeed. That faith, followed by the mentee producing high-quality science, will generate confidence.

Judging aptitude—can we?

Assessing aptitude has its own suite of challenges. Because of the intersection of social, psychological, experiential, and innate factors that affect our intellects and our ability to perform, it can be difficult to judge a student's ability to be a scientist. As mentors, it is our responsibility to examine the factors affecting a student's performance. A few questions we should ask include:

- Are my expectations reasonable for a scientist at this stage?

- Has this student had the training necessary to succeed at this task or in this environment (and could additional formal training improve their performance)?

- Does the student understand what is expected?

- Is this student disadvantaged in some way that makes the situation more difficult than it is for others?

- Is the student experiencing a stress—inside or outside the lab—that is affecting their performance?

- Might the student perform better in another environment?

Determining whether your expectations are clear and appropriate and whether a student has the necessary preparation can be accomplished through a dialogue with the student. The solutions to these issues should be agreed upon and implemented jointly. If the remedies do not result in satisfactory performance, then other actions may need to be taken.

Judging aptitude—impact of stress

People under stress cannot work at their highest potential; it may be impossible, therefore, to judge a stressed student's aptitude for science. Stress derives from many sources, some of which are obvious, some not so apparent The tension that we experience around deadlines is perceived by and understandable to most of us. But some students experience difficulties that may be invisible to us, and maybe even to the students themselves. Chronic illness and pain, financial problems, family responsibilities such as taking care of children or aging parents, or simply being different from the people around us can cause debilitating stress.

Some stress may come from past experience with prejudice. A student may worry that others will treat him differently if they find out that his parents are migrant farm workers, that he has epilepsy, or that he considered becoming a priest before choosing science. The student may have confronted bigotry in other situations that generated these fears and made him ultrasensitive to perceived or real intolerance. The student may be encountering prejudice in the lab that you may or may not perceive. There may be cliques from which he is excluded, jokes about his "difference" that may be intended to hurt him or are inadvertently hurtful. Discrimination experienced outside the lab or even off-campus might affect the student's ability to work. A person subjected to prejudice undergoes physiological changes in many different organ systems that translate into cognitive changes that influence the ability to focus, concentrate, and be creative. Even the fear or anticipation of such attitudes (known as "stereotype threat") can have crippling effects.

If you suspect that your student is suffering from stress that is affecting their ability to do science, consider discussing it with them. If the student has not discussed it with you, don't make assumptions or plunge in with aggressive questioning unless you know them very well and have established a trusting relationship. Instead, you can just provide an opening for the student to seize.

Session 4

55

QUESTIONABLE QUESTIONS (unless you have already developed a trusting relationship)	PROBABLY SAFE OPENERS
"Are you having marital problems?" "Did you break up with your girl-friend?"	"You seem a little down these days. Is everything OK?" "You're looking tired. I hope you're feeling OK."
"Are you spending too much time at the nursing home with your mother when you should be in lab?"	"Is your mother recovering from the stroke? (assuming the student had confided in you about the stroke)"
"What's it like to be a black man in this town, anyway?"	"I can imagine that being black in this very white environment might be difficult at times. If you ever want to talk about it, I'm here."
"It must be hard to explain what you do to your family with no col-lege graduates!"	"I was at a dinner with a bunch of lawyers the other night and, wow, did I struggle to explain what our lab does. Have you found any good analogies that lay people can relate to?"
"You're so attractive, you must get a lot of attention from the guys in the lab. Is it OK being the only woman on the 12th floor?"	"Are you comfortable in the lab? If there are ever conflicts, problems, or issues that get in the way of your work, will you please let me know what I can do to help?"
"Do you want to use my office dur-ing the day to pump milk while you're breastfeeding?"	"I can imagine that there are lots of logistical and practical issues that will arise when you have the baby. Please let me know if there is anything I can do to make things easier for you."
"Getting here for your graduation must be hard for your parents on a trash collector's salary, so do you want to use some of my frequent flyer miles to get them plane tickets?"	"I know you are counting on your parents being here for graduation. If there is anything I can do to help with their visit, let me know."

The inappropriate questions in the table are all intended to be kind and helpful, but may call attention to something that a student doesn't want singled out, causing embarrassment or awkwardness. If your students don't want to discuss their family, race, or nursing habits with you, respect that. The more appropriate questions attempt to pro-vide an opening that the student can take or decline. These questions express caring and show that you notice them as human beings, without intruding into private places where you might not be welcome.

Judging aptitude—innate ability

Many of us are frustrated that our students don't seem as smart as we think they should be. People mature intellectually at different rates and all of the factors discussed in the previous sections can affect apparent intelligence. It is also important to look around at people who have advanced in science and notice the characteristics that got them there. Some are simply brilliant, and the sheer power of their intellects has driven their success. But most have many other attributes that contributed to their success. Most highly successful scientists are extremely hard working, terrific managers and motivators of other people, colorful writers, and charismatic people. The fortunate (and often most successful) scientists have large doses of all of these traits, but many scientists have a mixture of strengths and weaknesses. Some are poor managers, others are unimpressive writers, and,

Case 1.

❝ I had an undergraduate student in my lab who didn't seem very bright and I doubted that he would make it as a scientist. I encouraged him to move on. The next time I saw him, he was receiving an award for outstanding undergraduate research that he did in another lab. I was surprised. The next time I encountered him was when I opened a top-notch journal and saw a paper with him as first author. I was impressed. Next I heard, he had received his PhD and was considered to be a hot prospect on the job market.

A couple of years later, I had a graduate student who was incredible bright and a wonderful person, but wasn't getting anything done. I had tried all of my mentoring tricks, and then borrowed some methods from others. In a fit of frustration, I encouraged the student to take a break from the lab and think about what to do next. While she was taking her break, she received an offer to complete her PhD in another lab. She did, published a number of highly regarded papers, landed a great postdoc, and is now a well-funded faculty member at a major research university.

These experiences have made me realize the power of the "match." The student, the lab, and the advisor have to be well matched, and all fit has to come together at the right time in the student's life. I can't be a good advisor to all students, and where I fail, someone else may succeed. It reminds me to be humble about mentoring, not to judge students, and never predict what they can't do. Happily, they will surprise you! ❞

amazingly, some don't seem all that smart or creative, yet their labs turn out great work because of their ability to create a highly effective research group.

There is room for lots of different kinds of people and intellects in science. A student who frustrates you with an excruciatingly linear or earthbound style of thinking may develop into a reliable and indispensable member of a research team. A student who can't seem to keep track of details in the lab may turn out to be a terrific professor who generates big ideas and relies on lab members to deal with the details. Before you judge a student, consider the diversity of people who make up the scientific community and ask yourself whether you can see your student being a contributor to that community. And ask yourself what each of those members of the community was like when they were at your student's stage of development.

Fairness: monitor prejudices and assumptions

Most of us harbor unconscious biases about other people that we apply to our evaluation of them. Few of us intend to be prejudiced, but culture and history shape us in ways that we don't recognize. Experiments show that people evaluate the quality of work differently if they are told that a man or a woman, a black or a white person performed the work (see "Benefits and Challenges of Diversity" in the next section for a detailed discussion of this research). We can't escape our culture and history, but we can try to hold ourselves to high standards of fairness and to challenge our own decisions. Regularly ask yourself if you would have reacted the same way to a behavior, a seminar, a piece of writing, or an idea if it was presented by someone of a different gender or race. When you evaluate people, make sure you are holding them all to the same standards. When you write letters of recommendation, check your language and content and make sure that you are not introducing subtle bias with the words you use or topics you discuss (see the next section for research on letters of recommendation for men and women).

Changing behavior

When we discover that a student is disorganized, introverted, or chronically late, what should we do? How much do we accommodate these differences to encourage diversity in our research community and when does accommodation become bad mentoring, hypocrisy, or a violation of the principles that we have agreed form our mentoring foundation? When is a behavior something that other students should tolerate and when does it violate the rights of others in the lab? These distinctions are tough to make, and we are likely to arrive at conclusions that differ from those of other mentors or even from our own judgments at other stages in our careers. Considering a few key questions may help clarify our mentoring decisions.

- Is the behavior creating an unsafe environment for the mentee or others in the lab?

- Is the behavior negatively affecting the productivity or comfort of others in the lab?

- Will the mentee be more effective, productive, or appreciated in the lab if the behavior or characteristic is modified?

- Is the behavior or characteristic sufficiently annoying to you that it interferes with your ability to work with the mentee?

Choose your battles carefully. If your answers to the questions are all "no," you may want to let the situation go. Sloppiness that creates a fire hazard or leads to poor data record-keeping must be corrected, but perhaps a desk strewn with papers, however irritating, can be ignored. A student who is introverted might be accommodated, but a student who is excessively talkative or boisterous and interfering with others' work needs to modify the behavior.

So, if a behavior needs to be changed, what's a mentor to do? If you are lucky, simply making the mentee aware of it may solve the problem. It will help to be directive about the type of change needed and why it is necessary. It is useful to lay out the problem that you are trying to solve and then ask the mentee to participate in developing the solution. If this doesn't work, you may need to use stronger language and eventually use sanctions to achieve the needed change.

Session 4

LESS EFFECTIVE	MORE EFFECTIVE
"Clean up your bench!"	"I'm concerned that the condition of your bench is creating a fire hazard. I'm sure you don't want to put the safety of the lab at risk, so what can we do to fix the situation?"
"Be on time to lab meetings from now on."	"You know, when you come into lab meeting fifteen minutes late, it's disruptive to the group and makes the person talking feel that their work isn't important to you. Is there some conflict in your schedule that I don't know about or do you think you can be on time in the future?"
"You'll never get anywhere in science if you don't dig in and stick with problems until you solve them."	"You seem to be giving up on solving this problem. I want to help you learn how to see problems through to their solutions, so what can I do to help? I want to work on this because problem-solving is going to be important throughout your career."

Case 2.

" Some issues are stickier than others. I once had a student who would come into the lab every Monday and loudly discuss his sexual exploits of the weekend. People in the lab—men and women—dreaded coming in on Mondays and were intensely uncomfortable during his discourse. No one in the group wanted to deal with it, and most of them were too embarrassed to even mention it to me. Finally, my trusted technician shared with me her intention to quit if this student didn't graduate very soon. I was faced with the challenge of telling the student that we all need to be sensitive to others in the lab and there might be people who didn't want to hear about his sex life.

I was uncomfortable with the conversation for a lot of reasons. First, I'm not used to talking to my students about their sex lives. Second, I was concerned that the student would be hurt and embarrassed that others in the lab had talked to me about his behavior and I didn't want to create a new problem in the process of solving the original one. Third, the student was gay and I didn't want him to think that his behavior was offensive because of this. I wanted him to appreciate that any discussion of sexual experience—straight or gay— was simply inappropriate for the open lab environment. But the student had never told me that he was gay, so I felt it was a further violation of his relationship with other lab members to indicate that I knew he was gay. The discussion did not go well because we were both so uncomfortable with the subject and I had trouble being as blunt as I should have been.

The behavior didn't change. The student finished his thesis and defended it. At the defense, one of the committee members suggested that the student do more experiments, and I detected the beginnings of a groundswell of support for his point of view. I blurted out that if this student stayed one more day in my lab, my wonderful technician would quit, so if he had to do more experiments, could he do them in one of their labs? In the end, everyone signed off on the thesis, the student graduated, and I never published the last chapter of the student's thesis because more experiments were needed to finish the story. I felt that I had weighed lab harmony against academic and scientific standards and have never been happy with how I handled the whole situation. "

Some behavior issues raise the questions of personal rights. Is it OK to rule that your students aren't allowed to wear headphones in the lab? That they dress a certain way? That they not put up posters or sayings that are offensive to others? That they aren't allowed to discuss politics or religion in the lab in ways that make some members uncomfortable? That they not make sexist or racist jokes? And whose definition of sexist and racist do we use? How do we balance overall lab happiness with the rights and needs of individuals?

Deciding what to do about problematic behavior may be one of the most annoying parts of being a mentor or lab leader. Many of us just wish everyone would know how to behave, get along, and get on with the science that we are here to do. Unfortunately, behavioral issues can prevent the science from getting done, and they just don't go away. Not dealing with some problems is unfair to the mentee, who deserves to know how he or she affects others, but the behavior must be addressed in a sensitive way to prevent embarrassment and animosity. Another question is, who should handle it? If you are a graduate student responsible for an undergraduate researcher, should you take care of the problem or ask your advisor to deal with it? If you are a lab leader, should you always deal with problems directly or is it sometimes appropriate to ask a member of the lab to tackle the problem diplomatically? These questions have to be answered in context and usually based on discussion with the other person who shares responsibility for the mentee.

Every mentoring relationship is different

Each person we mentor has their own unique set of needs and areas for growth. Use the beginning of the mentoring relationship to get to know your mentee and begin to experiment with ways of interacting. Does your mentee ask a lot of questions or do they need to be encouraged to ask more? Does your mentee respond well to direct criticism or do they need to be gently led to alternative answers or ways of doing things? In what areas do you think you can help your mentee the most—developing confidence, independence, and communication skills? Learning lab techniques and rigorous thinking? Improving interpersonal interactions? Does your mentee demand more time than you can or want to give, or do they need encouragement to seek you out more often? Mentoring relationships are as diverse as people, and they change over time. Monitor the relationship and make sure your mentoring style and habits are keeping up with the development of your mentee and the mentoring relationship.

As you assess progress in your mentoring relationship:
- Find your style—mentoring is personal and idiosyncratic.

- Communicate directly.

- Emphasize in your mentoring the aspects of science that are the most important—ethics, rigorous analytical thinking, risk-taking, creativity, and people.

- Be positive. Remember that people learn what quality is by having both the positive and the negative pointed out.

- Celebrate the differences among students.

- You are shaping the next generation—what do you want that generation to be?

Case 3.

" I am a graduate student in a very crowded lab. One summer, we hosted two students from Spain. The students were great—they worked hard, got interesting results, were fun to be around, and fit into the group really well. The problem was that they spoke Spanish to each other all day long. And I mean ALL DAY. For eight or nine hours every day, I listened to this loud rapid talking that I couldn't understand. Finally, one day I blew. I said in a not very friendly tone of voice that I'd really appreciate it if they would stop talking because I couldn't get any work done. Afterwards, I felt really bad and apologized to them. I brought the issue to my mentoring class and was surprised by the length of the discussion that resulted. People were really torn about whether it is OK to require everyone to speak in English and whether asking people not to talk in the lab is a violation of their rights. Our class happened to be visited that day by a Norwegian professor and we asked her what her lab policy is. She said everyone in her lab is required to speak in Norwegian. That made us all quiet because we could imagine how hard it would be for us to speak Norwegian all day long. "

An Important Mentor

" One of my most important mentors was Howard Temin. He had received the Nobel Prize a few years before I met him, but I didn't discover that until I had known him for a while and I never would have guessed, because he was so modest. Many aspects of science were far more important to Howard than his fame and recognition. One of those was young people. When he believed in a young scientist, he let them know it. As a graduate student, I served with Howard on a panel about the impact of industrial research on the university. It was the first time I had addressed a roomful of hundreds of people, including the press. My heart was pounding and my voice quavered throughout my opening remarks. I felt flustered and out of place. When I finished, Howard leaned over and whispered, "Nice job!" and flashed me the famous Temin smile. I have no idea whether I did a nice job or not, but his support made me feel that I had contributed something worthy and that I belonged in the discussion. I participated in the rest of the discussion with a steady voice.

When I was an assistant professor, I only saw Howard occasionally, but every time was memorable. One of the critical things he did for me—and for many other scientists—was to support risky research when no one else would. Grant panels sneered at my ideas (one called them "outlandish") and shook my faith in my abilities. Howard always reminded young scientists that virologists had resisted his ideas too, and reviews of his seminal paper describing the discovery of reverse transcriptase criticized the quality of the experiments and recommended that the paper be rejected! Howard was steadfast in his insistence that good scientists follow their instincts. When my outlandish idea turned out to be right, I paid a silent tribute to Howard Temin.

Howard showed support in many ways, some of them small but enormously meaningful. He was always interested in my work and often attended my seminars. When he was dying of cancer, his wife Rayla, a genetics professor, went home each day to make lunch for him. During that time, I gave a noon seminar on teaching that Rayla mentioned to Howard. When he heard who was giving the seminar, he told Rayla to attend it and that he would manage by himself that day. That was the last gift Howard gave me as a mentor before he died, and it will always live with me as the most important because it embodied everything I loved about Howard: he was selfless, generous, caring, and supportive.

At Howard's memorial service, students and colleagues spoke about how they benefited, as I had, from his enormous heart and the support that gave them the fortitude to take risks and fight difficult battles. Each of us who was touched by Howard knows that he left the world a magnificent body of science, but to us, his greatest legacy is held closely by the people who were lucky enough to have been changed by his great spirit. "

Session 5:
Resolving Challenges & Issues

Discussion Outline: Session 5

Topics:

Solutions to issues raised in case studies

How do you know things are going well with your mentee?

Case Study: Diversity

Midcourse Process Check

Describe Assignments for Session 6: Diversity

Materials for Mentors:

Case Studies: Diversity

Midcourse Process Check

Benefits and Challenges of Diversity

Session 5:

Discuss Solutions to Issues Raised in Case Studies

Some guiding questions:

1. *Did anyone find a specific solution helpful?*

2. *Has anyone tried one of the proposed solutions?*

3. *Did the presenter of the case study think the suggested solutions were feasible? Why or why not?*

Case Study: Diversity—Two cases

Many mentors find it challenging to work with students whose personalities differ from their own. Some find cultural differences awkward; some wonder whether their students experience a different lab environment from the one they experience. Some have never thought about any of these issues. The cases can be used to initiate a discussion on diversity.

Some guiding questions may include:

1. How do you deal with diversity?

2. How can you encourage different ways of thinking about science?

3. How can you accommodate different working styles?

4. What are some ways you can better understand your mentee's attitudes and experiences?

We have included an article entitled "Benefits and Challenges of Diversity" in this section.

Session 5

67

Midcourse Process Check

We recommend doing a midsemester process check to assess your discussion group. A sample survey form to conduct such a process check can be found in this section. Ideally, the responses from the mentors in your group will allow you to identify what is going well and what could use improvement during the second half of the seminar.

Assignment

Addressing diversity is complex. Ask the mentors in your group to consider the differences between them and their mentees at various levels. These can include differences in working style, ethnic differences, gender differences, differences in background, etc. Ask them to consider how these differences may influence their relationship and how they can learn from these differences.

Case Study: Diversity 1

❝ Last summer I worked with a fantastic undergraduate mentee. She was very intelligent and generated a fair amount of data directly relevant to my thesis project. I think that she had a positive summer research experience, but there are a few questions that still linger in my mind. This particular mentee was an African-American woman from a small town. I always wondered how she felt on a big urban campus. I also wondered how she felt about being the only African-American woman in our lab. In fact, she was the only African-American woman in our entire department that summer. I wanted to ask her how she felt, but I worried that it might be insensitive or politically incorrect to do so. I never asked. I still wonder how she felt and how those feelings may have affected her experience. ❞

Case Study: Diversity 2

❝ The biggest challenge I've encountered so far as a mentor was learning to work closely with someone whose personality and mannerisms are very different from my own. In my first interview with her, my student described herself as very laid-back and mentioned that she frustrates her parents with her "everything will take care of itself" attitude. This is a stark contrast to my personality and I find myself at times frustrated with her different work ethic. ❞

Mentoring Seminar Process Check

1. What is going well in this group?

2. What is not going so well in this group?

3. How do you feel about the structure, activities, and format of the group?

4. How do you feel about the topics we've discussed? What topics have we not considered that you would like to explore?

5. Additional comments:

Benefits and Challenges of Diversity

The diversity of the University's faculty, staff, and students influences its strength, productivity, and intellectual personality. Diversity of experience, age, physical ability, religion, race, ethnicity, gender, and many other attributes contributes to the richness of the environment for teaching and research. We also need diversity in discipline, intellectual outlook, cognitive style, and personality to offer students the breadth of ideas that constitutes a dynamic intellectual community.

Yet diversity of faculty, staff, and students also brings challenges. Increasing diversity can lead to less cohesiveness, less effective communication, increased anxiety, and greater discomfort for many members of a community (Cox 1993). To minimize the challenges and derive maximum benefits from diversity, we must be respectful of each other's cultural and stylistic differences and be aware of unconscious assumptions and behaviors that may influence interactions. The goal is to create a climate in which all individuals feel "personally safe, listened to, valued, and treated fairly and with respect" (Definition of Campus Climate, UW Provost's Office, 2004).

A vast and growing body of research provides evidence that a diverse student body, faculty, and staff benefits our joint missions of teaching and research.

Benefits for Teaching & Research

Research shows that diverse working groups are more productive, creative, and innovative than homogeneous groups. This research suggests that developing a diverse faculty will enhance teaching and research (Milem 2001). Some findings are:

- A controlled experimental study of performance in a brainstorming session compared the ideas generated by ethnically diverse groups composed of Asians, blacks, whites, and Latinos to those produced by ethnically homogenous groups composed of whites only. Evaluators who were unaware of the source of the ideas found no significant difference in the number of ideas generated by the two types of groups, but, using measures of feasibility and effectiveness, rated the ideas produced by diverse groups as being of higher quality (Cox 1993; McLeod, et al. 1996).

- The level of critical analysis of decisions and alternatives was higher in groups that heard minority viewpoints than in those that did not, regardless of whether or not the minority opinion was correct or ultimately prevailed. Minority viewpoints stimulated discussion of multiple perspectives and previously unconsidered alternatives (Nemeth 1985, 1995).

- A study of innovation in corporations found that the most innovative companies deliberately established diverse work teams (Kanter 1983).

- Using data from the 1995 Faculty Survey conducted by the Higher Education Research Institute (HERI) at UCLA, another study documented that scholars from minority groups have expanded and enriched scholarship and teaching in many intellectual disciplines by offering new perspectives, and raising new questions, challenges, and concerns (Antonio 2002; see also Turner 2000, Nelson and Pellet 1997).

- Several research studies found that women and faculty of color more frequently used active learning in the classroom, encouraged student input, and included perspectives of women and minorities in their course work (Milem 2001).

Benefits for Students:

Numerous research studies have examined the impact of diversity on students and educational outcomes. Cumulatively, these studies provide extensive evidence that diversity has a positive impact on all students, minority and majority (Smith et al. 1997). Some examples are:

- A national longitudinal study conducted by the Higher Educational Research Institute at UCLA involving 25,000 undergraduates attending 217 four-year colleges and universities in the late 1980s showed that institutional policies emphasizing diversity of the campus community, inclusion of themes relating to diversity in faculty research and teaching, and opportunities for students to confront racial and multicultural issues in the classroom and in extracurricular settings had uniformly positive effects on students' cognitive development, satisfaction with the college experience, and leadership abilities (Astin 1993).

- An analysis of two longitudinal studies, one using data from the Cooperative Institutional Research Program (CIRP), a national survey conducted by the Higher Educational Research Institute with more than 11,000 students from 184 institutions in 1985 and 1989, and one with approximately 1,500 students at the University of Michigan conducted in 1990 and 1994, showed that students who interacted with racially and ethnically diverse peers both informally and within the classroom showed the greatest "engagement in active thinking, growth in intellectual engagement and motivation, and growth in intellectual and academic skills" (Gurin 1999, Gurin et al. 2002).

- Another major study used data from the National Study of Student Learning (NSSL) to show that both in-class and out-of-class interactions and involvement with diverse peers fostered critical thinking. This study also showed a strong correlation between "the extent to which an institution's environment is perceived as racially nondiscriminatory" and students' willingness to accept both diversity and intellectual challenge (Pascarella et al. 1996).

- Using the "Faculty Classroom Diversity Questionnaire," a comprehensive survey of faculty attitudes toward and experiences with ethnic and racial diversity on campus, researchers found that more than 69% of approximately 500 faculty respondents in a randomly selected sample of 1,210 faculty from Carnegie Classified Research I institutions believed that all students benefited from learning in racially and ethnically diverse environments; that such environments exposed students to new perspectives and encouraged them to examine their own perspectives. More than 40% of respondents believed diversity fostered interactions that helped develop critical thinking and leadership skills (Maruyama and Moreno 2000). Another survey found that more than 90% of 55,000 faculty respondents believed that a racially and ethnically diverse campus enhanced students' educational experiences (Milem and Hakuta 2000).

- A 1993–94 survey of 1,215 faculty in doctoral-granting departments of computer science, chemistry, electrical engineering, microbiology,

Session 5

73

and physics showed that women faculty play an important role in fostering the education and success of women graduate students (Fox 2003).

Challenges of Diversity

Despite the benefits that a diversified faculty, staff, and student body offer to a campus, diversity also presents considerable challenges that must be addressed and overcome. Some examples include:

- Numerous studies show that women and minority faculty members are considerably less satisfied with many aspects of their jobs than are majority male faculty members. These include teaching and committee assignments, involvement in decision-making, professional relations with colleagues, promotion and tenure, and overall job satisfaction (Allen et al. 2002; Aguirre 2000; Astin and Cress 2003; Foster et al. 2000; Milem and Astin 1993; MIT Committee on Women Faculty 1999; Riger 1997; Somers 1998; Task Force on the Status of Women Faculty in the Natural Sciences and Engineering at Princeton 2003; Trower and Chait 2002; Turner 2002; Turner and Myers 2000; University of Michigan Faculty Work-Life Study Report 1999; Study of Faculty Worklife at the University of Wisconsin–Madison).

- A recent study of minority faculty in universities and colleges in eight Midwestern states (members of the Midwestern Higher Education Commission) showed that faculty of color experience exclusion, isolation, alienation, and racism in predominantly white universities (Turner and Myers, 2000).

- Minority students, as well, often feel isolated and unwelcome in predominantly white institutions and many experience discrimination and differential treatment. Minority status can result from race, ethnicity, national origin, sexual orientation, disability, and other factors (Amaury and Cabrera, 1996; Cress and Sax, 1998; Hurtado, 1999; Rankin, 1999; Smedley et al. 1993; Suarez-Balcazar et al. 2003).

- Women students, particularly when they are minorities in their classes, may experience "a chilly climate," which can include sexist use of language; presentation of stereotypic and/or disparaging views of women; differential treatment from professors; and sexual harassment (Crombie et al. 2003; Foster et al. 1994; Hall and Sandler 1982, 1984; Sands 1998; Swim et al. 2001; Van Roosmalen and McDaniel 1998; Sandler and Hall 1986; Whitte et al. 1999).

• Studies show that the lack of previous positive experiences with "outgroup members" (minorities) causes "ingroup members" (majority members) to feel anxious about interactions with minorities. This anxiety can cause majority members to respond with hostility or to simply avoid interactions with minorities (Plant and Devine 2003).

Influence of Unconscious Assumptions and Biases

Although we all like to think that we are objective scholars who judge people based entirely on merit and on the quality of their work and the nature of their achievements, copious research shows that every one of us brings with us a lifetime of experience and cultural history that shapes our interactions with others.

Studies show that people who have strong egalitarian values and believe that they are not biased may nevertheless unconsciously or inadvertently behave in discrimnatory ways (Dovidio 2001). A first step toward improving climate is to recognize that unconscious biases, attitudes, and other influences not related to the qualifications, contributions, behaviors, and personalities of our colleagues can influence our interactions, even if we are committed to egalitarian principles.

The results from controlled research studies in which people were asked to make judgments about human subjects demonstrate the potentially prejudicial nature of our many implicit or unconscious assumptions. Examples range from physical and social expectations or assumptions to those that have a clear connection to the environments in which we work.

Examples of Common Social Assumptions/Expectations:

• When shown photographs of people of the same height, evaluators overestimated the heights of male subjects and underestimated the heights of female subjects, even though a reference point, such as a doorway, was provided (Biernat and Manis 1991).

• When shown photographs of men with similar athletic abilities,

Session 5

evaluators rated the athletic ability of African-American men higher than that of white men (Biernat and Manis 1991).

- Students asked to choose counselors from among a group of applicants with marginal qualifications more often chose white candidates than African-American candidates with identical qualifications (Dovidio and Gaertner 2000).

These studies show how generalizations that may or may not be valid can be applied to the evaluation of individuals (Bielby and Baron 1986). In the study on height, evaluators applied the statistically accurate generalization that men are usually taller than women to their estimates of the height of individuals who did not necessarily conform to the generalization. If we can inaccurately apply generalizations to characteristics as objective and easily measured as height, what happens when the qualities we are evaluating are not as objective or as easily measured? What happens when, as in the studies of athletic ability and choice of counselor, the generalization is not valid? What happens when such generalizations unconsciously influence the ways we interact with other people?

Examples of assumptions or biases that can influence interactions:

- When rating the quality of verbal skills as indicated by vocabulary definitions, evaluators rated the skills lower if they were told that an African-American provided the definitions than if they were told that a white person provided them (Biernat and Manis 1991).

- When asked to assess the contribution of skill and luck to successful performance of a task, evaluators more frequently attributed success to skill for males and to luck for females, even though males and females performed the task identically (Deaux and Emswiller 1974).

- Evaluators who were busy, distracted by other tasks, and under time pressure gave women lower ratings than men for the same written evaluation of job performance. Gender bias decreased when they gave ample time and attention to their judgments, which rarely occurs in actual work settings (Martell 1991).

- Evidence suggests that perceived incongruities between the female gender role and leadership roles create two types of disadvantage for women: (1) ideas about the female gender role cause women to be perceived as having less leadership ability than men and consequently impede women's rise to leadership positions, and (2) women in leadership positions receive less favorable evaluations because they are perceived to be violating gender norms. These perceived incongruities lead to attitudes that are less positive toward female leaders than male leaders (Eagly and Karau 2002; Ridgeway 2001).

- A study of the nonverbal responses of white interviewers to black and white interviewees showed that white interviewers maintained higher levels of visual contact, reflecting greater attraction, intimacy, and respect, when talking with whites, and higher rates of blinking, indicating greater negative arousal and tension, when talking with blacks (Dovidio et al. 1997).

Examples of assumptions or biases in academic contexts:

Several research studies have shown that biases and assumptions can affect the evaluation and hiring of candidates for academic positions. These studies show that assessment of résumés and postdoctoral applications, evaluation of journal articles, and the language and structure of letters of recommendation are significantly influenced by the gender of the person being evaluated. As we attempt to enhance campus and department climate, we need to consider whether the influence of such biases and assumptions also affects selection of invited speakers, conference participants, interaction and collaboration with colleagues, and promotion to tenure and full professorships.

- A study of over 300 recommendation letters for medical faculty hired at a large American medical school in the 1990s found that letters for female applicants differed systematically from those for males (Trix and Psenka 2002).

- In a national study, 238 academic psychologists (118 male, 120 female) evaluated a résumé randomly assigned a male or a female name. Both male and female participants gave the male applicant better evaluations for teaching, research, and service and were more likely to hire the male than the female applicant (Steinpreis et al. 1999).

- A study of postdoctoral fellowships awarded by the Medical Research Council in Sweden found that women candidates needed substantially more publications to achieve the same rating

Session 5

as men, unless they personally knew someone on the panel (Wenneras and Wold 1997).

- In a replication of a 1968 study, researchers manipulated the name of the author of an academic article, assigning a name that was male, female, or neutral (initials). The 360 college students who evaluated this article were influenced by the name of the author, evaluating the article more favorably when it was written by a male than when written by a female. Questions asked after the evaluation was complete showed that bias against women was stronger when evaluators believed that the author identified only by initials was female (Paludi and Bauer 1983).

Biases and assumptions can influence women, minorities, and the university in the following ways:

- Women and minorities may be subject to higher expectations in areas such as number and quality of publications, name recognition, or personal acquaintance with a committee member.

- Colleagues from institutions other than the major research universities that have trained most of our faculty may be undervalued. Opportunities to benefit from the experiences and expertise of colleagues from other institutions, such as historically black universities, four-year colleges, government, or industry, who can offer innovative, diverse, and valuable perspectives on research, teaching, and the functioning of the department, may consequently be neglected.

- The work, ideas, and findings of women or minorities may be undervalued, or unfairly attributed to a research director or to collaborators despite contrary evidence in publications or letters of reference.

- The ability of women or minorities to run a research group, raise funds, and supervise students and staff may be underestimated, and may influence committee and teaching assignments.

- Assumptions about possible family responsibilities and their effect on a colleague's career path may negatively influence evaluation of merit, despite evidence of productivity, and may affect committee and teaching assignments.

- Negative assumptions about whether female or minority colleagues "fit in" to the existing environment can influence interactions.

References

Amaury, Nora and Alberto F. Cabrera. 1996. The role of perceptions of prejudice and discrimination on the adjustment of minority students to college. *The Journal of Higher Education* 67:119-148.

Antonio, Anthony. 2002. Faculty of color reconsidered: Reassessing contributions to scholarship. *Journal of Higher Education* 73:582-602.

Astin, Alexander W. 1993a. Diversity and multiculturalism on the campus: How are students affected? *Change* 25:44-50.

Astin, Alexander W. 1993b. *What matters in college? Four critical years revisited.* San Francisco, CA: Jossey-Bass, Inc., Publishers.

Bielby, William T. and James N. Baron. 1986. Sex segregation and statistical discrimination. *American Journal of Sociology* 91:759-799.

Biernat, Monica and Melvin Manis. 1991. Shifting standards and stereotype-based judgments. *Journal of Personality and Social Psychology* 66:5-20.

Cox, Taylor H. Jr. 1993. *Cultural diversity in organizations: Theory, research and practice.* San Francisco: Berrett-Koehler Publishers.

Cress, Christine M. and Linda J. Sax. Summer 1998. Campus climate issues to consider for the next decade. *New Directions for Institutional Research* 98:65-80.

Crombie, Gail et al. 2003. Students' perceptions of their classroom participation and instructor as a function of gender and context. *Journal of Higher Education* 74:51-76.

Deaux, Kay and Tim Emswiller. 1974. Explanations of successful performance on sex-linked tasks: What is skill for the male is luck for the female. *Journal of Personality and Social Psychology* 29:80-85.

Dovidio, John F. 2001. On contemporary prejudice: The third wave. *Journal of Social Issues* 57:829-849.

Dovidio, John F. and S.L. Gaertner. 2000. Aversive racism and selection decisions: 1989 and 1999. *Psychological Science* 11:315-319.

Dovidio, John F. et al. 1997. The nature of prejudice: Automatic and controlled processes. *Journal of Experimental Social Psychology* 33:510-540.

Eagly, Alice H. and Steven J. Karau. July 2002. Role congruity theory of prejudice toward female leaders. *Psychological Review* 109:573-597.

Foster, Sharon W. et al. June 2000. Results of a gender-climate and work-environment survey at a midwestern academic health center. *Academic Medicine* 75:653-60.

Foster, Ted J. et al. April 1994. An empirical test of Hall and Sandler's 1982 Report: Who finds the classroom climate chilly? Paper presented at the annual meeting of the Central States Communication Association. Oklahoma City, OK.

Fox, Mary Frank. 2003. Gender, faculty, and doctoral education in science and engineering. In *Equal rites, unequal outcomes: Women in American research universities.* Ed. Lilli S. Hornig. New York: Kluwer Academic.

Session 5

79

Gurin, Patricia. 1999. Expert Report of Patricia Gurin, *Gratz, et al. v. Bollinger, et al.*, No. 97-75321 (E.D. Mich.) & *Grutter, et al. v. Bollinger, et al.*, No. 97-75928 (E.D. Mich.). In *The compelling need for diversity in higher education. (http://www.umich.edu/~urel/admissions/legal/expert/qual.html)*

Gurin, Patricia et al. 2002. Diversity and higher education: Theory and impact on educational outcomes. *Harvard Educational Review* 72:330-366.

Hall, Roberta M. and Bernice R. Sandler. 1982. *The classroom climate: A chilly one for women?* Washington, D.C.: Association of American Colleges, Project on the Status and Education of Women.

Hall, Roberta M. and Bernice R. Sandler. 1984. *Out of the classroom: A chilly campus climate for women.* Washington, D.C.: Association of American Colleges, Project on the Status and Education of Women.

Heilman, Madeline E. 1980. The impact of situational factors on personnel decisions concerning women: Varying the sex composition of the applicant pool. *Organizational Behavior and Human Performance* 26:386-395.

Hurtado, Sylvia et al. 1999. Enacting diverse learning environments: Improving the climate for racial/ethnic diversity. *ASHE-ERIC Higher Education Reports* 26; 8:1-116.

Maruyama, Geoffrey and José F. Moreno. 2000. University faculty views about the value of diversity on campus and in the classroom. In *Does diversity make a difference? Three research studies on diversity in college classrooms.* Washington, D.C.: American Council on Education and American Association of University Professors, 9-35.

Martell, Richard F. 1991. Sex bias at work: The effects of attentional and memory demands on performance ratings for men and women. *Journal of Applied Social Psychology* 21: 1939-60.

McLeod, Poppy L., Sharon A. Lobel, and Taylor H. Cox. 1996. Ethnic diversity and creativity in small groups. *Small Group Research* 27: 248-265.

Milem, Jeffrey F. and Kenji Hakuta. 2000. The benefits of racial and ethnic diversity in higher education. In *Minorities in higher education: Seventeenth annual status report.* Ed. D. Wilds. Washington, D.C.: American Council on Education.

Milem, Jeffrey F. 2003. The Educational benefits of diversity: Evidence from multiple sectors. In *Compelling interest: Examining the evidence on racial dynamics in higher education.* Ed. Mitchell Chang et al. Stanford: Stanford Education.

MIT Committee on Women Faculty. 1999. A study on the status of women faculty in science at MIT. Boston: Massachusetts Institute of Technology. *(http://web.mit.edu/fnl/women/women.html#The%20Study)*

Nelson, Stanley and Gail Pellet. 1997. *Shattering the silences [videorecording].* San Francisco: Gail Pellet Productions. California Newsreel.

Nemeth, Charlan J. 1985. Dissent, group process, and creativity: The contribution of minority influence. *Advances in Group Processes* 2:57-75.

Nemeth, Charlan J. 1995. Dissent as driving cognition, attitudes, and judgments. *Social Cognition* 13:273-291.

Paludi, Michele A. and William D. Bauer. January 1983. Goldberg revisited: What's in an author's name. *Sex Roles* 9; 1:387-90.

Pascarella, Ernest T. et al. 1996. Influences on students' openness to diversity and challenge in the first year of college. *Journal of Higher Education* 67:174-196.

Plant, E. Ashby and Patricia G. Devine. 2003. The antecedents and implications of interracial anxiety. *Personality and Social Psychology Bulletin* 29:790-801.

Rankin, Susan. 1999. Queering campus: Understanding and transforming climate. *Metropolitan Universities: An International Forum* 9; 4:29-38.

Ridgeway, Cecilia L. 2001. Gender, status, and leadership. *Journal of Social Issues* 57:637-655.

Riger, Stephanie et al. 1997. Measuring perceptions of the work environment for female faculty. *Review of Higher Education* 21; 1:63-78.

Sandler, Bernice R. and Roberta M. Hall. 1986. *The campus climate revisited: Chilly for women faculty, administrators, and graduate students.* Washington, D.C.: Association of American Colleges, Project on the Status and Education of Women.

Sands, Roberta G. 1998. Gender and the perception of diversity and intimidation among university students. *Sex Roles* 39:801-815.

Smith Daryl G. et al. 1997. *Diversity works: The emerging picture of how students benefit.* Washington D.C.: Association of American Colleges and Universities.

Somers, Patricia et al. 1990. Faculty and staff: The weather radar of campus climate. *New Directions for Institutional Research* 98 (Summer 1998):35-52.

Steinpreis, Rhea, Katie A. Anders, and Dawn Ritzke. 1999. The impact of gender on the review of the curricula vitae of job applicants and tenure candidates: A national empirical study. *Sex Roles* 41:509-528.

Study of faculty work-life at the University of Wisconsin–Madison. Unpublished Report.

Suarez-Balcazar, Yolanda et al. 2003. Experiences of differential treatment among college students of color. *Journal of Higher Education* 74: 428-444.

Swim, Janet K. et al. Spring 2001. Everyday sexism: evidence for its incidence, nature, and psychological impact from three daily diary studies. *Journal of Social Issues* 57:31-53.

Trix, Frances and Carolyn Psenka. 2003. Exploring the color of glass: Letters of recommendation for female and male medical faculty. *Discourse & Society* 14:191-220.

Trower, Cathy A. and Richard P. Chait. March 2002–April 2002. Faculty diversity: Too little for too long. *Harvard Magazine* 33-37, 98.

Turner, Caroline S. V. Sept/Oct 2000. New faces, new knowledge. *Academe* 86:34-37.

Turner, Caroline S. V. 2002. *Diversifying the faculty: A guidebook for search committees.* Washington, D.C.: Association of American Colleges and Universities.

Turner, Caroline S. V. Jan/Feb 2002. Women of color in academe. *The Journal of Higher Education* 73:74-93.

Turner, Carolyn S.V. and Samuel L. Myers. 2000. *Faculty of color in academe: bittersweet success.* Boston: Allyn and Bacon.

University of Michigan Work-life study report. November 1999. *(http://www.umich.edu/%7Ecew/fwlsreport.pdf)*

Van Roosmalen, Erica and Susan A. McDaniel. 1998. Sexual harrassment in academia: A hazard to women's health. *Woman and Health* 28:33.

Wenneras, Christine and Agnes Wold. 1997. Nepotism and sexism in peer-review. *Nature* 387:41-43.

Whitte, Elizabeth et al. 1999. Women's perceptions of a 'chilly climate' and cognitive outcome in college: Additional evidence. *Journal of College Student Development* 20:163-177.

Session 5

Evelyn Fine and Jo Handelsman, Women In Science and Engineering Leadership Institute (WISELI), University of Wisconsin–Madison. http://wiseli.engr.wisc.edu

81

Session 6:
Evaluating Our Progress as Mentors

Discussion Outline: Session 6

Topics:

Discuss Mentoring Challenges and Solutions.

Discussion Questions
- How do you and your mentee differ?

Describe Assignment for Session 7: Present Mentoring
Challenge to Adviser

Session 6

Session 6

Mentoring Challenges and Solutions

Have mentors continue to share stories about mentoring. Open discussion questions might include:

1. *What is the biggest challenge you are facing as a mentor?*
2. *What has been your biggest success as a mentor?*
3. *What has been your biggest disappointment as a mentor?*

How do you and your mentee differ?

In the previous session, the mentors were asked to think about how they differ from their mentee and how those differences may influence both the attitude of the mentee as well as the attitude of the mentor towards the mentee. You may wish to revisit this topic in this session, especially if some of the case studies raise issues of diversity.

Assignment

Ask mentors to present mentoring challenges to their research adviser or another person they respect as a mentor and ask that person how they would handle the situation. This assignment opens lines of communication between graduate student/postdoctoral mentors and their advisers on a topic that they both can relate to: mentoring. We have also asked students to write about their adviser's reactions to the challenge and reflect on the response.

Session 7:
The Elements of Good Mentoring

Discussion Outline: Session 7

Topics:

Discussion Questions:

- What can we learn from the other mentors around us?

- What has proven effective in your mentoring?

Preparing Students for Presentations

Describe Assignment for Session 8: Rewriting Your Mentoring Philosophy.

Materials for Mentors:

Mentor-Mentee Check-in Questions: Ensuring an Excellent Finish

"Righting Writing"

Discussion 7:

What can we learn from other mentors around us?

This discussion can focus on the conversations the mentors had with their own advisers. Some suggested questions for discussion are:

1. Did you find discussing a mentoring challenge with your adviser to be useful?

2. In general, do you think it is useful to discuss mentoring challenges with your colleagues who are mentors themselves?

3. Did your discussion with your adviser alter your perception of their role as a mentor?

4. Are you more likely to discuss issues of mentoring with your adviser/other colleagues in the future?

What has proven effective in your mentoring?

Have the mentors identify effective tools to include in a "mentoring toolbox." Some of these might be:

- Identifying the mentees' goals

- Evaluating our mentees' understanding

- Evaluating our mentees' talents and building on them

- Developing a relationship founded on mutual respect

- Giving mentees' ownership of their work and promoting accountability

- Sharing our own experiences

- Creating the interactive lab environment

- Identifying what motivates each mentee

Session 7

- Balancing belief with action and experience

- Creating a safe environment in which mentees feel that it is acceptable to·fail and learn from their mistakes

- Encouraging growth through challenges

- Promoting learning through questioning

- Walking experimental avenues together

Presentations

At this point in many research programs, many mentees are preparing to present their work in a talk, poster, or paper. Some guiding questions for mentors and mentees to consider when planning final presentations are included in this section. You may ask the mentors to generate a list of guidelines for presentation preparation. These might include:

- Think about simplicity.

- Think about clarity.

- Focus on the big picture. What do they want people to remember about the project? Why should the audience care?

- Start preparing early!

- Spend time introducing the students to the technology of the computer programs they will need to generate the poster if they have not used them before.

- Have the student think about their audience. Ask them what they like to hear or see in a presentation.

- Have your student practice their presentation more than once:

  a. Practice for other members of your lab or department.
  b. Do not let the first practice be for the principle investigator.
  c. Students for whom English is a second language may want/need more practice; be sensitive to this.

- Decide together on the starting material. Will they have access to your text, data, figures, etc., or are they building their own from scratch? What would they prefer?

- Be helpful and constructive, but remember this is their presentation, NOT yours.

We have included an article entitled "Righting Writing" in this section that may be useful in improving scientific writing.

Assignment

Ask the mentors to rewrite their mentoring philosophies. It is important for each mentor to reflect on their original philosophy to determine whether they were in fact able to practice their philosophy and whether their philosophy has changed.

Ensuring an Excellent Finish

Goals

- Reaffirm the expectations of both mentor and mentee.

- Assess the progress you have made in completing your research project.

- Determine what can reasonably be accomplished in the remainder of the program.

- Outline a strategy for completing the final paper and preparing the final presentation.

Students:

- Do you feel that you are achieving the goals you outlined at the beginning? Why or why not?

- What do you believe has been your greatest accomplishment in the laboratory so far?

- What has been the most frustrating part about working in the laboratory? How can your mentor help you deal with this?

- How do you feel about the progress you have made on your research project thus far?

- What would you still like to accomplish with regard to your research project?

- How can your mentor help you in writing your final paper and in preparing your final presentation?

- How would you like to maintain contact with your mentor once the program has ended? Would you like to ask your mentor to write you a letter of recommendation in the future for your application to graduate or professional school, or for employment?

Session 7

Mentors:

- What do you think has been your mentee's greatest accomplishment in the laboratory so far?

- What have you learned from mentoring this student in your laboratory?

- How do you feel about the progress you and your mentee have made on this research project?

- What would you still like your mentee to accomplish with regard to the research project?

- How can you help your student in writing their final paper and in preparing their final presentation?

- How would you like to maintain contact with your student once the program has ended? Would you be willing to write a letter of recommendation for your student in the future?

These guidelines were developed by Janet Branchaw, Center for Biology Education, University of Wisconsin, based on Zachary, L.J. (2000). *The Mentor's Guide: Facilitating Effective Learning Relationships.* San Francisco, CA: Jossey-Bass, Inc., Publishers.

Session 7

Righting Writing

•

Jo Handelsman

Getting Started

Perhaps the most important aspects of writing occur before you even put pen to paper or fingertips to keyboard. First, try to answer the following questions for yourself:

- Who is my audience?
- Why should they care?
- What are my major points?

Who is my audience?

The answer to the first question will help you define how you start your paper, the angle you take in presenting the significance of the work, and the background information you need to supply. For example, if you are presenting the discovery of a plasmid in a pathogen of trees in a forestry journal, you might emphasize the importance of the pathogen to forestry and provide background information about plasmids and their importance, but you can assume that people reading the journal understand the significance of trees. However, a paper presenting the same discovery in a journal on plasmids would not need to discuss the significance of plasmids, but the audience would not be expected to know much about trees and their pathogens.

Why should they care?

You must catch a reader's attention. Everyone has more than enough to read these days, and most people will just turn the page of a journal if the first paragraph of a paper does not capture their imagination or make a compelling case for the paper's significance. When you start writing, assume that your reader is uninterested in your topic and it is your challenge to make it interesting. You may use its significance to society or its relevance to solving practical, human problems, or you may use the pure intellectual interest of an unsolved biological problem or a paradox that needs to be explained. Whatever your angle, make it clear, concise, and honest. Usually, what interests you about the project will be interesting to your readers.

What are my major points?

Most people learn new information best when it is presented in small bits organized around an interesting concept. If you bombard your audience with too many new ideas, they are unlikely to understand them all well. If you focus your paper around one or two key ideas, it will be more cohesive and cleanly structured. Therefore, before you start writing, choose your most important points. If a reader were to learn only one thing from your paper, what would you like it to be?

The Global Issues

Good writing is typified by clear, flowing organization. In an organized piece, the reader's mind moves easily from one idea to the next through the writer's effective use of connections, transitions, and logical organization. Below are a few suggestions to help you develop the overall logic and organization of your writing.

1.　The Lead

The first sentence of a piece of writing is critical. It clues the reader in to the central theme and catches attention. This is particularly important in a personal statement associated with an application to graduate school or for a job. Make your first sentence interesting, but not too long or complex—you don't want the reader to get tired on the first sentence. Be sure that your word use and grammar are absolutely correct. There is nothing as damning in an application as a glaring grammatical error in the first sentence of a personal statement. Finally, be sure that the rest of your piece lives up to the first sentence. Don't tease the reader with a neat idea and then fail to develop it.

2.　Organization

Make an outline. Justify to yourself why each section should be included. What is its relationship to your topic, theme, or hypothesis? Identify the essential information and then try to streamline your material, but be thorough. It is better to review a smaller amount of information thoughtfully than to cover a great volume superficially.

Use paragraphs and subheadings to provide the reader with a sense of the organization of concepts. Lead or topic sentences can help define the content of each paragraph for the reader, but be careful not to simply repeat a subheading in the first sentence of the paragraph.

Session 7

95

3. Transitions

Try to make explicit connections between sentences, paragraphs, and sections. Avoid lists of ideas or sentences that are not connected. Remembering this rule will make your writing more fluid, force you to make mental connections between ideas, and motivate your audience to read further. Reading a list of unconnected ideas often makes a reader say, "So what?" Logical connections will lead a reader to say, "Oh, I see!"

Compare the following paragraphs:

- Genetic diversity is a powerful tool in biotechnology. Many strains of bacteria have been used for production of vinegar, antibiotics, and enzymes in industrial microbiology. Crop varieties adapted to many different environments are used in agriculture.

- Throughout the history of biotechnology, genetic diversity has been a powerful tool. In microbiology, genetically diverse strains of bacteria have been used to maximize production of vinegar, antibiotics, and enzymes. In agriculture, genetic diversity has been exploited to produce crop varieties adapted to many different environments.

The first paragraph is a list of apparently unrelated pieces of information. The second paragraph connects the three sentences. The similar construction of the second and third sentences, starting with "In microbiology," and "In agriculture," provides a signal to the reader that these are examples of the point made in the topic sentence. This is reinforced by the use of the phrases "genetically diverse" and "genetic diversity" in the second and third sentences, indicating that they are examples of the overall concept of genetic diversity in biotechnology

Specific Issues

The construction of each sentence is critical to enhance the clarity and impact of writing. While the specifics may seem picky or unimportant, the most minor mistakes can make your writing ambiguous, boring, or hard to read. Below are some pointers that will help your writing be comprehensible and interesting.

1.   Stacked modifiers

In writing about science, we have a tendency to use strings of adjectives, or stacked modifiers, to avoid lots of prepositional phrases. Use of extra words is usually discouraged, but they can be very welcome to readers if they help you avoid dense sentences filled with many stacked modifiers. This is especially important for readers who are not familiar with the jargon of your field. It is often hard for a reader who is unfamiliar with the material to figure out how the words in a series of stacked modifiers fit together. An example is "cryptic plasmid subclones." Is the plasmid cryptic, or are the subclones cryptic? This would be clearer if it was written: "subclones of the cryptic plasmid." Two short words have been added, but the ambiguity is gone.

2.   Hyphens

Another way to handle stacked modifiers is to hyphenate the modifiers to distinguish them from the noun. An example is "weak root pathogen." If the pathogen affects weak roots, it should read "weak-root pathogen." (The alternative meaning is a root pathogen that is weak; in this case, do not hyphenate). Do not hyphenate two-word descriptors when one of the words is an adverb. (These can usually be spotted by their "-ly" endings.) For example, "genetically engineered microorganisms" and "randomly generated mutants" should not be hyphenated.

3.   Verbs

Follow the usual rules. What is published or generally known is presented in the present tense, your results are presented in the past tense, and predicted results should be in the conditional tense. A common mistake is to describe predicted results in the past tense, and this can make it very hard for the reader to distinguish between what happened and what might happen.

Avoid the passive voice. Never use the wordy passive.

Active voice:	The plants grew rapidly.
Passive:	Rapid growth of the plants was observed.
Wordy passive:	It was observed that plant growth was rapid.

Verbs provide the spice in scientific writing. Search for interesting, active verbs to stimulate your reader's imagination. Examine the phrases from writing by Paul Ehrlich,

Session 7

one of the most persuasive writers on the topic of biodiversity conservation.

> "The food resource...in all major ecosystems is the energy that green plants **bind** into organic molecules...."

> "...our species can safely **commandeer** upwards of 80% of...."

> "**Arresting** the loss of diversity will be extremely difficult."

> "...the **spewing** of toxins into the environment..."

> from *Biodiversity*, ed. E.O. Wilson

6. Word use

Variety. Try not to use different nouns for the same subject. Many students purposely interchange "bacterium," "cell," and "organism" for variety. This can be very confusing to the reader. Science writing is precise, and no two words mean the same thing, so consistently use the one that is appropriate for your meaning.

Pretentious and empty words. Try to avoid pretentious words that can be replaced by simple, direct words. Some examples:

there exists (there is)
by means of (with, by)
utilize (use)
due to the fact that (since, because)
in order to (to)

Try to cut all words that do not advance your ideas. "Empty" words are those that slow down the reader and obscure meaning. An example is: "experiments proposed in this investigation will..." In this phrase, "in this investigation" adds nothing. They are empty words. Weeding out empty words makes your writing more vigorous and direct.

To find empty words, focus on the main point of the sentence. Identify the subject and the verb. Where is the action in the sentence? Identify the words that contribute to that idea and delete phrases that add nothing.

Waffle words. Use sparingly and avoid more than one in a sentence.

Excessive waffling:	The data may suggest that the bacteria could swim.
Really excessive waffling:	The data may potentially suggest that the bacteria might be able to swim.
Just the right amount of waffling:	The data suggest that the bacteria swim.
Excessive waffling:	It appears that the plasmid may potentially transfer to other bacteria.

Just the right amount of waffling: The plasmid may transfer to other bacteria.

Latin names. Match your verbs properly to Latin word endings.

Singular: The bacterium is fast.

Plural: The bacteria are fast.

When the genus name is turned into a colloquial name, don't capitalize it: "rhizobia," "pseudomonads," "enterococci," "bacilli."

7.   Writing in parallel

To save words and achieve maximum clarity, use the same grammatical structure in two parts of a compound sentence. If you change verb tense in the middle of the sentence, the second part tends to dangle.

Nonparallel: Plants require water for root growth and producing seed.
Parallel: Plants require water for root growth and seed production.
Parallel: Plants require water to produce roots and seed.

Nonparallel: Seed exudate may inhibit growth of beneficial bacteria and suppressing infection of seeds by pathogens.
Parallel: Seed exudate may inhibit growth of beneficial bacteria and suppress infection of seeds by pathogens.

If you include a list of items, try to start each member of the list with the same form of speech. For example, study the following list of objectives:

Nonparallel: To clone the gene.
Sequencing the gene.
The function of the gene will be determined.

Parallel: To clone the gene.
To sequence the gene.
To determine the function of the gene.

8.   The dreaded "that" vs. "which"

The words "that" and "which" have different uses in English, although they are often used interchangeably. The following rule used to be followed strictly in all good writing, but many people ignore it now. It is still useful, and adhering to it makes writing less ambiguous.

"That" is used in restrictive clauses, and "which" is used in nonrestrictive clauses,

Session 7

which are usually preceded by a comma. This may sound trivial, but the differences in meaning can be significant. Look at the following:

> The pestagon that generates research about insects is in Davis, CA.
> The Pestagon, which generates research about insects, is in Davis, CA.

In the first sentence, the dependent clause is "that generates research about insects" and it is absolutely essential to the sentence. It defines which pestagon, in a group of pestagons, the sentence is about. From this sentence, we infer that there must be other pestagons, but the one the writer is telling us about is the one that generates research about insects.

In the second sentence, the independent clause, "which generates research about insects," is incidental. It is an aside that tells a reader something about the Pestagon, but does not distinguish this pestagon from other pestagons. The implication of this sentence is that there is only one Pestagon.

If you can't remember the rule about clauses, look for the comma. A comma always precedes the appropriate use of "which" in the middle of a sentence.

Drafting a Scientific Paper

Readers usually expect a scientific paper to adhere to the following organization:

Title. Use a concise phrase that captures the most important point of the paper.

Abstract. Provide your reader with a synopsis of your work that will stand alone and stimulate the reader's curiosity.

Introduction. The introduction identifies the topic in broad terms to capture the widest diversity of readers, offers a specific illustration of that topic, then explains why the topic is important and articulates a research question or claim that your paper will answer.

Methods. Provide a clear description of what you did and how you did it, complete enough that someone could repeat your experiments successfully. Reference published methods appropriately. The balance between how much you say and how much you reference depends on the journal.

Results. Introduce the results with a brief rationale (no more than a phrase or a sentence) for what you have done. Then launch into what you have found. Don't simply repeat what is obvious in the tables and figures. Restate only the most important results, and then use this section to indicate patterns and trends in the data.

Figures and Tables. Make sure the legends are clear and complete, enabling a reader to make sense of your findings without reading the text. Make sure that headings of columns in tables refer to the data and units in the columns. Make sure that axes on graphs are labeled clearly and units are defined.

Discussion. Start with a summary of the important findings in your paper, drawing them together in a new way that doesn't simply repeat the Results section. Then launch into interpretation. Why are your data significant, and what new insight do they give us into your research question? Into your topic in general? Do they point us in new directions or promote a new understanding of an old concept? How do your results articulate with previous findings in your field? What cautions must we use in interpreting or extrapolating these results and what limitations are intrinsic to your methods? Finally, what are the next key steps, how does your work lay the foundation for them, and how will they contribute to the larger picture of the field?

Remember that the less you say, the greater the impact of what you do say. Be absolutely ruthless with the Discussion—make a list of the points you want your reader to understand and then write a paragraph about each. If you go on and on, your readers will lose your key points and you are probably restating results or delving into obscure detail. If you must elaborate, make it absolutely clear to your reader why all of these points must be considered. The Discussion of a paper is often the most difficult and fun to write. This is where you craft your science, giving it emphasis, texture, and context.

All the paragraphs in the Intro, Results, and Discussion should be connected with transitions that explain how the concept you've just finished writing about relates to what you're about to start discussing. Use outlines, topic sentences, and key concepts to structure your text. If it's not clear to you what you want to say before you write it, you can be sure your readers won't get it.

—With contributions from Christina Matta and Brad Hughes

Session 7

Session 8:
Developing a
Mentoring Philosophy

Discussion Outline: Session 8

Topics:

Sharing Mentoring Philosophies

Question and Answer Time about the Seminar and/or Mentoring

Have Participants Complete an Evaluation

Session 8

Sharing Mentoring Philosophies

Make certain that each mentor receives copies of all the mentoring philosophies well before the session. Discuss the mentoring philosophies both collectively and individually. Some guiding questions may include:

- What are common themes in the mentoring philosophies?

- What is memorable about certain philosophies that might stand out to a search committee, for example?

- Has your perception of mentoring changed during the semester? If so, how?

- What was your original definition of a mentor? Has your definition changed? If so, how?

- How would you approach your mentoring differently next time?

Questions?

Leave time for the students to ask questions before leaving the group. You may want to leave time for the members of the group to share what they found effective in the mentor training and what they would like to see changed. Conduct an evaluation during this session or after the last session (see Evaluation section).

Evaluation of the Mentoring Seminar

Evaluation Protocols

Mentee Survey

Twenty-seven questions covering the mentee's research experiences and their mentor, increase in skill and knowledge areas, and career goals.

Deployed one week before completing their research experiences.

Mentor Survey

Twenty-six questions covering their mentee and their experience mentoring, increase in skill levels of both the mentee and mentor, previous mentoring experience, career goals, and experience in mentor training.

Deployed one week after the mentees have completed their research experiences.

Facilitator Survey

Eighteen questions covering their mentee and their experience mentoring, increase in skill levels of both the mentee and mentor, previous mentoring experience, career goals, and experience in mentor training.

Deployed within one week after the seminar is completed.

Survey content and questions:
Christine Pfund, cepfund@wisc.edu
Copyright © 2004 by the Board of Regents
of the University of Wisconsin System

Survey format and deployment:
Zoomerang: www.zoomerang.com
Copyright ©1999–2004 by MarketTools, Inc.

Mentoring Seminar: Mentee Survey

1. My current grade level:

   ___ 1st year undergraduate   ___ 4th + year undergraduate

   ___ 2nd year undergraduate   ___ 1st year graduate student

   ___ 3rd year undergraduate   ___ 2nd + year graduate student

   ___ 4th year undergraduate   ___ Postdoctoral researcher

   ___ Faculty member

   ___ Other, please specify:_____

2. The institution at which I conducted my research was:

   _____

3. The department in which I conducted my research was:

   _____

4. Which type of program did you participate in, if any, in conjunction with the research experience?

   ___ No program (individual research)

   ___ Independent research as part of a course

   ___ Academic year undergraduate research program

   ___ Summer research program

   ___ Lab rotation

5. How long did you work on this research project?

   ___ 1–3 weeks

   ___ 4–7 weeks

   ___ 8–10 weeks

   ___ 11–13 weeks

___ 14–16 weeks

___ Other, please specify (in weeks):_____

6. Please describe the types of activities you engaged in as a researcher (e.g., collected field data, analyzed computer data, cloned DNA, observed animal behavior, etc.).

7. Please use the following scale to identify your skill level in the following areas BEFORE your research experience and NOW:

1   No Skill
2   Very Low Skill
3   Low Skill
4   Moderate Skill
5   High Skill
6   Very High Skill

Understanding scientific papers: BEFORE

1___   2___   3___   4___   5___   6___

Understanding scientific papers: NOW

1___   2___   3___   4___   5___   6___

Using research equipment: BEFORE

1___   2___   3___   4___   5___   6___

Using research equipment: NOW

1___   2___   3___   4___   5___   6___

Formulating research hypotheses: BEFORE

1___   2___   3___   4___   5___   6___

Formulating research hypotheses: NOW

1___   2___   3___   4___   5___   6___

Developing a research project: BEFORE

1___   2___   3___   4___   5___   6___

Developing a research project: NOW

1___   2___   3___   4___   5___   6___

1	No Skill
2	Very Low Skill
3	Low Skill
4	Moderate Skill
5	High Skill
6	Very High Skill

Conducting a research project: BEFORE

1___   2___   3___   4___   5___   6___

Conducting a research project: NOW

1___   2___   3___   4___   5___   6___

Analyzing data: BEFORE

1___   2___   3___   4___   5___   6___

Analyzing data: NOW

1___   2___   3___   4___   5___   6___

8.   **Continued**...Please use the following scale to identify your skill level in the following areas BEFORE your research experience and NOW.

1   No Skill
2   Very Low Skill
3   Low Skill
4   Moderate Skill
5   High Skill
6   Very High Skill

Giving feedback to a peer: BEFORE

1___   2___   3___   4___   5___   6___

Giving feedback to a peer: NOW

1___   2___   3___   4___   5___   6___

Receiving feedback: BEFORE

1___   2___   3___   4___   5___   6___

Receiving feedback: NOW

1___   2___   3___   4___   5___   6___

Presenting information: BEFORE

1___   2___   3___   4___   5___   6___

Presenting information: NOW

1___   2___   3___   4___   5___   6___

Articulating questions: BEFORE

1___   2___   3___   4___   5___   6___

Articulating questions: NOW

1___   2___   3___   4___   5___   6___

Dealing with setbacks: BEFORE

1___   2___   3___   4___   5___   6___

Dealing with setbacks: NOW

1___   2___   3___   4___   5___   6___

Working independently on research: BEFORE

1___   2___   3___   4___   5___   6___

Working independently on research: NOW

1___   2___   3___   4___   5___   6___

Working collaboratively with others: BEFORE

1___   2___   3___   4___   5___   6___

Working collaboratively with others: NOW

1___   2___   3___   4___   5___   6___

Your research skills, in general: BEFORE

1___   2___   3___   4___   5___   6___

Your research skills, in general: NOW

1___   2___   3___   4___   5___   6___

1	No Skill
2	Very Low Skill
3	Low Skill
4	Moderate Skill
5	High Skill
6	Very High Skill

9. Please use the following scale to identify your level of knowledge in the following areas BEFORE your research experience and NOW.

    1   No knowledge
    2   Very little knowledge
    3   Little knowledge
    4   Some knowledge
    5   Much knowledge

    The nature of science and research: BEFORE

    1___   2___   3___   4___   5___

    The nature of science and research: NOW

    1___   2___   3___   4___   5___

    The nature of the job as a researcher: BEFORE

    1___   2___   3___   4___   5___

    The nature of the job as a researcher: NOW

    1___   2___   3___   4___   5___

    Career paths of science faculty: BEFORE

    1___   2___   3___   4___   5___

    Career paths of science faculty: NOW

    1___   2___   3___   4___   5___

    What graduate school is like: BEFORE

    1___   2___   3___   4___   5___

    What graduate school is like: NOW

    1___   2___   3___   4___   5___

    Career options in the sciences, in general: BEFORE

    1___   2___   3___   4___   5___

    Career options in the sciences, in general: NOW

    1___   2___   3___   4___   5___   6___

    In the following questions, **MENTOR** is defined as the person who was assigned to work with you on research, or who was responsible for providing direction to you, supervising you, helping you, answering your questions, signing off on assignments, etc.

10.   My primary mentor is a(n):

____   Undergraduate student

____   Graduate student

____   Postdoc

____   Scientist or lab technician

____   Faculty member

____   Other, please specify:_____

11.   On average, how many hours per week did you spend working on the research project?

____   0–5

____   6–10

____   11–20

____   21–30

____   31–40

____   More than 40

12.   On average, how many hours per week did you spend in face-to-face contact with your mentor?

____   0–5

____   6–10

____   11–20

____   21–30

____   31–40

____   More than 40

13.   The amount of time my mentor spent with me was:

____   Too little

____   Just right

____   Too much

14. Please respond to the statements below using the following scale, regarding your primary mentor:

1   **My mentor did not do this.**
2   **My mentor tried to do this, but was ineffective.**
3   **My mentor did this sometimes, and was effective.**
4   **My mentor did this frequently, and was effective.**

My mentor gave me an overview of how my research fit into an overall research project.

1___   2___   3___   4___

My mentor helped me develop my research skills.

1___   2___   3___   4___

My mentor showed interest in my research project.

1___   2___   3___   4___

My mentor was available to me when I had problems or questions about my research.

1___   2___   3___   4___

My mentor offered constructive feedback when necessary.

1___   2___   3___   4___

My mentor and I developed a relationship based on trust.

1___   2___   3___   4___

My mentor understood how I learn best.

1___   2___   3___   4___

My mentor created an environment that allowed me to achieve my goals.

1___   2___   3___   4___

My mentor seemed so busy that I was afraid to interrupt her/him.

1___   2___   3___   4___

My mentor had an effective mentoring style.

1___   2___   3___   4___

My mentor acted as a positive role model.

1___   2___   3___   4___

15. Please respond to the statements below using the following scale, regarding your primary mentor:

1   My mentor did not do this.
2   My mentor tried to do this, but was ineffective
3   My mentor did this sometimes, and was effective.
4   My mentor did this frequently, and was effective.

My mentor showed interest in me as a person.

1___   2___   3___   4___

My mentor fostered my independence.

1___   2___   3___   4___

My mentor encouraged me to have confidence in my skills.

1___   2___   3___   4___

My mentor appreciated my contributions.

1___   2___   3___   4___

My mentor encouraged me to be creative.

1___   2___   3___   4___

My mentor made me enthusiastic about my project.

1___   2___   3___   4___

My mentor helped me feel curious about my project.

1___   2___   3___   4___

My mentor treated me as a colleague.

1___   2___   3___   4___

My mentor helped me decide on a career path.

1___   2___   3___   4___

My mentor communicated her/his expectations of me.

1___   2___   3___   4___

My mentor respected my goals.

1___   2___   3___   4___

My mentor allowed me to take ownership in my research.

1___   2___   3___   4___

My mentor created an environment where I felt safe to make mistakes.

1___   2___   3___   4___

My mentor made me feel included in the lab.

1___   2___   3___   4___

My mentor regularly assessed skills and knowledge that I gained in the lab.

1___   2___   3___   4___

16.   What are the characteristics that made your mentor effective?

17.   What are the characteristics that made your mentor less effective than s/he could be?

18.   During your research experience, how often did your mentor ask for feedback about her/his mentoring style and effectiveness?

___   Never

___   Once

___   Twice

___   Three times

___   Four times

___   Weekly

___   Other, please specify:_____

19.   Would you recommend your mentor to another student researcher?

           ____ YES   ____ NO

Please explain:

20.   Have you had a "mentored" research experience previous to this one?

           ____ YES   ____ NO

If yes, how did this experience compare to your previous one?

21.   How much time did you spend with the professor who runs the research group, if s/he was not your primary mentor?

    ___   Not applicable; the professor was my primary mentor

    ___   Never

    ___   1–2 times during the entire experience

    ___   Monthly

    ___   Weekly

    ___   1–2 times per week

    ___   More than twice per week

    ___   Daily

22.   What was the single most important thing you learned during your research experience?

23.   What was the best part of your research experience?

24.   Would you like to continue doing research either with the research group from this project or with another group?

        ____YES   ____NO

25.   Which of the following do you plan to apply to as a result of your research experience? Check all that apply.

    ____   Graduate school

    ____   Medical school

    ____   Professional school

    ____   Scholarship/fellowship

    ____   I do not plan to apply for any of the above.

    ____   Other, please specify:_____

26.   Do you plan to pursue a career in the sciences as a result of your experience in the lab?

        ____ YES   ____ NO

27.   Is there anything else you would like to share about your research experience (including suggestions about what to keep and what to change about the program)?

A Mentoring Seminar: Mentor Survey

1. I am a(n):

   ___ Undergraduate student

   ___ Graduate student

   ___ Postdoc

   ___ Lab technician

   ___ Scientist

   ___ Faculty member, nontenured

   ___ Faculty member, tenured

   ___ Other, please specify: _____

2. My primary department is:

   _____

3. My university is:

   _____

4. I served as the primary mentor to a student researcher. (**Mentor** is defined as the person who is primarily responsible for providing direction and guidance to the student researcher ["**mentee**"]).

   ____YES   ____NO

5. Did you participate in a mentoring seminar while mentoring a student researcher? If not, go to question 9.

   ____YES   ____NO

6. The name of the person who facilitated my mentoring seminar was:

   _____

7.   How valuable was the discussion of each topic?

    1   **Not applicable**
    2   **Not useful or interesting**
    3   **Interesting, but not useful**
    4   **Useful, but not interesting**
    5   **Useful and interesting**

Elements of a good research project

1___   2___   3___   4___   5___

Establishing a good relationship with mentee

1___   2___   3___   4___   5___

Designing research projects

1___   2___   3___   4___   5___

Setting goals and establishing expectations

1___   2___   3___   4___   5___

Sharing mentoring challenges with each other

1___   2___   3___   4___   5___

Designing approaches to address mentoring challenges

1___   2___   3___   4___   5___

Addressing issues of diversity

1___   2___   3___   4___   5___

Evaluating your own progress as a mentor

1___   2___   3___   4___   5___

Articulating a mentoring philosophy

1___   2___   3___   4___   5___

Applying scientific teaching to your mentoring

1___   2___   3___   4___   5___

8.   Would you recommend the mentoring seminar to a colleague?

    ____YES   ____NO

Please explain:

9.   Which of the following did you do as a mentor? **Check all that apply**.

___   Designed your mentee's study before his/her arrival

___   Ordered necessary supplies before his/her arrival

___   Discussed goals and outcomes of your mentee's research project

___   Discussed expectations of your mentee with him/her

___   Discussed your mentee's expectations of you, as a mentor

___   Discussed amount of time s/he would spend on the research project per week

___   Oriented your mentee to your lab/facility and its practices

___   Oriented your mentee to your building/facility

___   Introduced your mentee to others in the lab/facility and its practices

___   Talked with your mentee about things other than research

___   Discussed career goals with your mentee

___   Discussed scientific papers with your mentee

___   Reflected upon or wrote your own mentoring philosophy

___   Discussed mentoring issues with your advisor

___   Discussed mentoring issues with other colleagues

___   Considered issues of diversity related to mentoring

___   Other, please specify:_____

10.   Using the following scale, please indicate the level at which you engaged in each of the following mentoring objectives:

1   **This is not one of my mentoring objectives.**
2   **I have considered how to include this in my mentoring.**
3   **I have tried to do this in my mentoring.**
4   **I have evidence that I do this effectively in my mentoring.**

Provide mentee with authentic research experience

1___   2___   3___   4___

Help mentee develop research skills

1___   2___   3___   4___

1	This is not one of my mentoring objectives.
2	I have considered how to include this in my mentoring.
3	I have tried to do this in my mentoring.
4	I have evidence that I do this effectively in my mentoring.

Help mentee decide about a career path and science

1___   2___   3___   4___

Build mentee's confidence

1___   2___   3___   4___

Foster mentee's independence

1___   2___   3___   4___

Foster open communication

1___   2___   3___   4___

Determine whether mentee understands me

1___   2___   3___   4___

Build a relationship with mentee based on trust

1___   2___   3___   4___

Set reasonable goals for project

1___   2___   3___   4___

Create an environment where mentee can achieve goals

1___   2___   3___   4___

Stimulate mentee's creativity

1___   2___   3___   4___

Stimulate mentee's curiosity

1___   2___   3___   4___

Choose mentoring strategies consistent with my philosophy

1___   2___   3___   4___

Consult my colleagues for advice on mentoring

1___   2___   3___   4___

Reflect on the effectiveness of my mentoring strategies

1___   2___   3___   4___

Apply scientific teaching to my mentoring

1___   2___   3___   4___

Devise and implement diverse solutions to mentoring challenges

1___   2___   3___   4___

11.   Which is the most challenging aspect of providing a mentee with a research experience? Choose one.

___   Assessing mentee's background (knowledge and skills)

___   Dealing with mentee's inexperience (knowledge and skill)

___   Keeping mentees engaged

___   Allocating time

___   Finding resources

___   Identifying mentee's motivations

___   Remaining patient

___   Addressing mentee's misconceptions about science

___   Setting reasonable goals for project

___   Building mentee's confidence

___   Fostering mentee's independence

___   Deciding on the "best solution" to a given mentoring challenge

___   Other, please specify:_____

12.   Using the scale below, please identify YOUR skill level **BEFORE** working with your mentee and **NOW**.

1   No skill
2   Very low skill
3   Low skill
4   Moderate skill
5   High skill
6   Very high skill

Helping a student plan a research project: BEFORE

1___   2___   3___   4___   5___   6___

Helping a student plan a research project: NOW

1___   2___   3___   4___   5___   6___

Assessing a student's learning and understanding: BEFORE

1___   2___   3___   4___   5___   6___

Assessing a student's learning and understanding: NOW

1___   2___   3___   4___   5___   6___

Building a student's confidence: BEFORE

1___   2___   3___   4___   5___   6___

Building a student's confidence: NOW

1___   2___   3___   4___   5___   6___

Giving a student feedback: BEFORE

1___   2___   3___   4___   5___   6___

Giving a student feedback: NOW

1___   2___   3___   4___   5___   6___

Developing strategies to deal with mentoring challenges: BEFORE

1___   2___   3___   4___   5___   6___

Developing strategies to deal with mentoring challenges: NOW

1___   2___   3___   4___   5___   6___

Helping a student prepare a paper, presentation, or a poster: BEFORE

1___   2___   3___   4___   5___   6___

Helping a student prepare a paper, presentation, or a poster: NOW

1___   2___   3___   4___   5___   6___

1	No skill
2	Very low skill
3	Low skill
4	Moderate skill
5	High skill
6	Very high skill

Building mentee's independence: BEFORE

1___   2___   3___   4___   5___   6___

Building mentee's independence: NOW

1___   2___   3___   4___   5___   6___

Establishing expectations: BEFORE

1___   2___   3___   4___   5___   6___

Establishing expectations: NOW

1___   2___   3___   4___   5___   6___

Building a relationship based on trust and respect: BEFORE

1___   2___   3___   4___   5___   6___

Building a relationship based on trust and respect:NOW

1___   2___   3___   4___   5___   6___

Addressing diversity issues: BEFORE

1___   2___   3___   4___   5___   6___

Addressing diversity issues: NOW

1___   2___   3___   4___   5___   6___

Consulting colleagues to help solve mentoring challenges: BEFORE

1___   2___   3___   4___   5___   6___

Consulting colleagues to help solve mentoring challenges: NOW

1___   2___   3___   4___   5___   6___

Resolving conflicts in mentoring: BEFORE

1___   2___   3___   4___   5___   6___

Resolving conflicts in mentoring: NOW

1___   2___   3___   4___   5___   6___

13. I feel that the overall quality of the mentoring I provided was:

    ___ Excellent

    ___ Good

    ___ Fair

    ___ Poor

14. What academic level was your mentee?

    ___ 1st year undergraduate

    ___ 2nd year undergraduate

    ___ 3rd year undergraduate

    ___ 4th year undergraduate

    ___ 4th + year undergraduate

    ___ 1st year graduate student

    ___ 2nd + year graduate student

    ___ Postdoctoral researcher

    ___ Faculty member

    ___ Other, please specify:_____

15. In which type of program did your mentee participate, in conjunction with the research experience? **Check all that apply.**

    ___ No program (individual research)

    ___ Independent research as part of a course

    ___ Academic year undergraduate research program

    ___ Summer research program for undergraduates

    ___ Lab rotation

    ___ Other, please specify:_____

16. Prior to this research experience, my mentee worked with our research team for:

    ___ No time; this was the mentee's first experience with our team

    ___ Less than one semester

_____ 1 year

_____ 2 years

_____ More than 2 years

_____ Other, please specify:_____

17.   On average, approximately how many hours per week did your mentee work on research?

_____ 0–5

_____ 6–10

_____ 11–20

_____ 21–30

_____ 31–40

_____ More than 40

18.   On average, approximately how many hours per week did you spend in face-to-face contact with your mentee work?

_____ 0–5

_____ 6–10

_____ 11–20

_____ 21–30

_____ 31–40

_____ More than 40

19.   The amount of time spent with your mentee was:

_____ Too little

_____ Just right

_____ Too much

20.   Using the scale below, please identify your **MENTEE'S** skill level in the following areas **BEFORE** his/her research experience and **NOW**:

1   No Skill
2   Very Low Skill
3   Low Skill
4   Moderate Skill
5   High Skill
6   Very High Skill

Understanding scientific papers: **BEFORE**

1___   2___   3___   4___   5___   6___

Understanding scientific papers: **NOW**

1___   2___   3___   4___   5___   6___

Using lab equipment: **BEFORE**

1___   2___   3___   4___   5___   6___

Using lab equipment: **NOW**

1___   2___   3___   4___   5___   6___

Formulating research hypotheses: **BEFORE**

1___   2___   3___   4___   5___   6___

Formulating research hypotheses: **NOW**

1___   2___   3___   4___   5___   6___

Developing a research project: **BEFORE**

1___   2___   3___   4___   5___   6___

Developing a research project: **NOW**

1___   2___   3___   4___   5___   6___

Conducting a research project: **BEFORE**

1___   2___   3___   4___   5___   6___

Conducting a research project: **NOW**

1___   2___   3___   4___   5___   6___

Analyzing data: **BEFORE**

1___   2___   3___   4___   5___   6___

Analyzing data: **NOW**

1___ 2___ 3___ 4___ 5___ 6___

Giving feedback to a peer: **BEFORE**

1___ 2___ 3___ 4___ 5___ 6___

Giving feedback to a peer: **NOW**

1___ 2___ 3___ 4___ 5___ 6___

Receiving feedback: **BEFORE**

1___ 2___ 3___ 4___ 5___ 6___

Receiving feedback: **NOW**

1___ 2___ 3___ 4___ 5___ 6___

Presenting information: **BEFORE**

1___ 2___ 3___ 4___ 5___ 6___

Presenting information: **NOW**

1___ 2___ 3___ 4___ 5___ 6___

Articulating questions: **BEFORE**

1___ 2___ 3___ 4___ 5___ 6___

Articulating questions: **NOW**

1___ 2___ 3___ 4___ 5___ 6___

Dealing with setbacks: **BEFORE**

1___ 2___ 3___ 4___ 5___ 6___

Dealing with setbacks: **NOW**

1___ 2___ 3___ 4___ 5___ 6___

Working independently: **BEFORE**

1___ 2___ 3___ 4___ 5___ 6___

Working independently: **NOW**

1___ 2___ 3___ 4___ 5___ 6___

Working collaboratively with others: **BEFORE**

1___ 2___ 3___ 4___ 5___ 6___

1	No Skill
2	Very Low Skill
3	Low Skill
4	Moderate Skill
5	High Skill
6	Very High Skill

Working collaboratively with others: **NOW**

1___   2___   3___   4___   5___   6___

His/Her research skills, in general: **BEFORE**

1___   2___   3___   4___   5___   6___

His/Her research skills, in general: **NOW**

1___   2___   3___   4___   5___   6___

21.   I feel that the overall quality of my mentee's performance was:

___   Excellent

___   Good

___   Fair

___   Poor

22.   Had you mentored an undergraduate in a research setting before this semester?

___YES   ___NO

If yes, what did you do differently in the two experiences?

23.   Would you mentor an undergraduate researcher again?

___YES   ___NO

What would you do differently if you were to mentor again?

24.   Overall, was being a mentor a positive experience?

        ____YES   ____NO

      Why or why not?

25.   Have your career goals changed as a result of your experience as
      a mentor?

        ____YES   ____NO

      Please explain:

26.   Has this mentoring experience changed your view of your own mentor
      (e.g., your advisor/PI of the research team)?

        ____YES   ____NO

      Please explain:

      Comments:

A Mentoring Seminar: Facilitator Survey

1.   I am a:

   ___   Graduate student

   ___   Postdoc

   ___   Lab technician

   ___   Scientist

   ___   Faculty member, nontenured

   ___   Faculty member, tenured

   ___   Other, please specify:_____

2.   My primary department is:

   _____

3.   My university is

   _____

4.   My role in the mentoring seminar was:

   ___   Facilitator (I taught the seminar by myself.)

   ___   Cofacilitator (I taught the seminar with at least one other person.)

5.   The number of students in the mentoring seminar was: _____

6.   Please indicate how the mentors in your seminar were identified and recruited.

7. Using the scale below, please rate how valuable the following elements were in the mentoring seminar?

    1   Not used in the seminar
    2   Not useful or interesting
    3   Interesting, but not useful
    4   Useful, but not interesting
    5   Useful and interesting

    The "Entering Mentoring" Manual overall

    1___   2___   3___   4___   5___

    The big questions in mentoring

    1___   2___   3___   4___   5___

    The guiding questions

    1___   2___   3___   4___   5___

    The facilitator notes

    1___   2___   3___   4___   5___

    The case studies

    1___   2___   3___   4___   5___

    The readings

    1___   2___   3___   4___   5___

    The assignments

    1___   2___   3___   4___   5___

    The discussions

    1___   2___   3___   4___   5___

8. Please comment on what you found particularly valuable in any of the above categories.

9. Using the following scale, please indicate how valuable the discussion of each topic was in your mentoring seminar:

    1   **We did not discuss this topic.**
    2   **Not useful or interesting**
    3   **Interesting, but not useful**
    4   **Useful, but not interesting**
    5   **Useful and interesting**

    Elements of a good research project

    1___   2___   3___   4___   5___

    Establishing a good relationship with mentee

    1___   2___   3___   4___   5___

    Designing research projects

    1___   2___   3___   4___   5___

    Setting goals and establishing expectations

    1___   2___   3___   4___   5___

    Sharing mentoring challenges with each other

    1___   2___   3___   4___   5___

    Designing approaches to address mentoring challenges

    1___   2___   3___   4___   5___

    Addressing issues of diversity

    1___   2___   3___   4___   5___

    Evaluating your own progress as a mentor

    1___   2___   3___   4___   5___

    Articulating a mentoring philosophy

    1___   2___   3___   4___   5___

    Applying scientific teaching to your mentoring

    1___   2___   3___   4___   5___

10. Please comment on what you found particularly valuable in any of the preceeding discussions.

11. Based on the discussion in your seminar, which do you feel mentors find to be the most challenging aspect of providing a mentee with a research experience? **Choose one.**

   ___ Assessing mentee's background (knowledge and skills)

   ___ Dealing with mentee's inexperience (knowledge and skill)

   ___ Keeping mentees engaged

   ___ Allocating time

   ___ Finding resources

   ___ Identifying mentee's motivations

   ___ Remaining patient

   ___ Addressing mentee's misconceptions about science

   ___ Setting reasonable goals for project

   ___ Building mentee's confidence

   ___ Fostering mentee's independence

   ___ Deciding on the "best solution" to a given mentoring challenge

   ___ Other, please specify:_____

12. Which of the following do you think the majority of your mentors did? **Check all that apply.**

   ___ Designed their mentee's study before his/her arrival

   ___ Ordered necessary supplies before his/her arrival

   ___ Discussed goals and outcomes of their mentee's research project

   ___ Discussed expectations of their mentee with him/her

   ___ Discussed their mentee's expectations of them, as a mentor

_____ Discussed amount of time s/he would spend on the research project per week

_____ Oriented their mentee to their lab/facility and its practices

_____ Oriented their mentee to their building/facility

_____ Introduced their mentee to others in the lab/facility and its practices

_____ Talked with their mentee about things other than research

_____ Discussed career goals with their mentee

_____ Discussed scientific papers with their mentee

_____ Reflected upon or wrote their own mentoring philosophy

_____ Discussed mentoring issues with their advisor

_____ Discussed mentoring issues with other colleagues

_____ Considered issues of diversity related to mentoring

_____ Other, please specify:_____

13. Overall, the quality of the mentors in the mentoring seminar was:

_____ Excellent

_____ Good

_____ Fair

_____ Poor

14. I feel that the overall quality of the facilitation I provided was:

_____ Excellent

_____ Good

_____ Fair

_____ Poor

15. Would you facilitate the mentoring seminar again:

_____YES   _____NO

What would you do differently if you were to facilitate the seminar again?

16.   Overall, was being a facilitator a positive experience?

        ____YES   ____NO

Why or why not?

17.   Would you recommend facilitating the mentoring seminar to a colleague?

        ____YES   ____NO

Please explain:

18.   Has your own philosophy of mentoring changed as a result of your experience as a facilitator of the mentoring seminar?

        ____YES   ____NO

Please explain:

Comments:

Resources

Adams, H.G. 1992. *Mentoring: An essential factor in the doctoral process for minority students.* National Center for Graduate Education for Minorities in Engineering and Science, Inc. (GEM).

Barker, Kathy. 2002. *At the helm: A laboratory navigator.* Cold Spring Harbor Laboratory Press, Cold Spring Harbor, New York.

Handelsman, J. 2003. Teaching scientists to teach. In *The Howard Hughes Medical Institute Bulletin 31.(http://www.hhmi.org/bulletin/june2003/)*

Handelsman, J. et al. 2004. Scientific Teaching. *Science* 304:51-52.

Howard Hughes Medical Institute and Burroughs Welcome Fund. 2004. Making the right moves: A practical guide to scientific management for postdocs and new faculty.

Mabrouk, P.A. 2003. Research learning contracts: A useful tool for facilitating successful undergraduate research experiences. *Council on Undergraduate Research Quarterly* 24: 26-30.

Merkel, C.A. and S.M. Baker. 2002. *How to mentor undergraduate researchers.* Council on Undergraduate Research. *(http://www.cur.org/)*

Monte, Aaron. 2001. Mentor expectations and student responsibilities in undergraduate research. *Council on Undergraduate Research Quarterly* 21:66-71.

National Academy of Sciences, National Academy of Engineering, and Institute of Medicine. 1997. *Adviser, teacher, role model, friend: On being a mentor to students in science and engineering.* Washington, D.C.: National Academy Press.

National Institutes of Health, Office of the Director. 2002. A guide to training and mentoring in the intramural research program at NIH. Bethesda, MD: National Institutes of Health. *(http://www1.od.nih.gov/oir/sourcebook/ethic-conduct/TrainingMentoringGuide_7.3.02.pdf)*

National Research Council of the National Academies. 2003. *BIO 2010.* Washington D.C.: National Academies Press.

Seymour, E., A.B. Hunter, S.L. Laursen, and T. Deantoni. 2004. Establishing the benefits of research experiences for undergraduates in the sciences: First findings from a three-year study. *Science Education* 88; 4:493-534.

Shea, G.F. 2000. *Mentoring.* Menlo Park, CA.: Crisp Publications.

Shellito, C., K. Shea, G. Weissmann, A. Mueller-Solger and W. Davis. 2001. Successful mentoring of undergraduate researchers: Tips for creating positive student research experiences. *Journal of College Science Teaching* 30:460-464.

University of Michigan, Horace H. Rackham School of Graduate Studies. 2002. How to mentor graduate students: A guide for faculty at a diverse university. Ann Arbor, MI: University of Michigan Press. *(http://www.rackham.umich.edu/StudentInfo/Publications/FacultyMentoring/contents.html)*

Vasgird, Daniel and Ellen Hyman-Browne. *Responsible conduct of research: mentoring.* *(http://www.ccnmtl.columbia.edu/projects/rcr/rcr_mentoring/)*

Zachary, L.J. 2000. *The mentor's guide: Facilitating effective learning relationships.* San Francisco, CA: Jossey-Bass, Inc., Publishers.

Appendix

"No Dumb Questions" Seminar:

Enriching the Research Experience for
Undergraduates in Science

Overview of the "No Dumb Questions" Seminar

Goal

The goal of the "No Dumb Questions" seminar is to enrich the research experience for undergraduates, to help them learn about what it means to be a scientist, and to build their confidence in science research.

Format

The weekly seminar provides a forum for open discussion among undergraduates and between undergraduates and faculty about topics that aren't addressed in other settings. The central feature of this seminar is that students are made to feel safe so they ask questions that they feel embarrassed to ask in other settings. Based on their questions, students can learn more about the scientific method, deepen their understanding of the research discipline, gain exposure to other scientific disciplines, practice oral and written communication skills, receive advice about graduate school and career choices, and develop a network of peers and support that expands beyond the research lab.

The strategy is to require students to ask questions that they consider dumb. When undergraduates are presented with the challenge of asking a "dumb question," they typically find it humorous, but quickly learn to use the opportunity to ask questions that they are afraid or embarrassed to ask elsewhere. They discover that there really is no such thing as a dumb question—that questions they thought were silly or trivial, or had answers that were obvious to everyone else, are often astute and generate good discussion. The range of questions that students ask includes: "What is agar made of?" "How do you decide who is an author on a paper?" "Can someone please explain homologous recombination? My mentor has explained it six times and he'll think I'm an idiot if I ask again." "How much does it cost to go to graduate school?" When the students engage in this forum, they often show visible relief and gratitude for the chance to ask questions that have preoccupied them. The further discovery that their questions are considered complex, intriguing, and worthwhile by experienced scientists reveals layers of the practice and culture of science that they often miss in other environments.

Implementation

Logistics: The seminar can be done easily in weekly one-hour sessions. At the University of Wisconsin–Madison, we have found that 8-12 undergraduate students is ideal, but the seminar can be done well with more or fewer students, or with high school students. If possible, teach the seminar in a small, informal classroom that is conducive to group work. A conference or meeting room typically works better than a classroom with fixed desks or lab tables.

Time frame: This seminar is most effective during an intensive summer research program where students work full-time in a research lab on an independent project. It could also be implemented during the academic year as part of an independent research project or a for-credit course.

Getting participants: Who participates is up to you. It's most important that you start with a cohort of people who can dedicate time to the entire seminar series. At the University of Wisconsin–Madison, we have found that more diverse groups of students offer the most rewarding discussions. For example, senior undergraduates with prior lab experience can provide support for younger students, and students from the home institution can help visiting students with logistics. The more heterogeneous the group's experience, the better the discussions.

Scientific teaching: As an instructor, use this teaching experience for your own teaching portfolio, whether you are being reviewed for tenure, seeking a new job, or simply hoping to gain teaching experience as part of your career. Use the evaluation forms to gather information from the students' experience, and use it as evidence of your teaching.

Institutional support and future directions: Let your department chair or campus administrators know that you are doing this course, and follow up with a report of the participant evaluations. We always find teaching evaluations from the "No Dumb Questions" seminar to be very high, and they can be used to encourage campus-wide adoption of the seminar.